Complete Audio Mastering

Practical Techniques

Gebre Waddell

Mc
Graw
Hill
Education

New York   Chicago   San Francisco
Lisbon   London   Madrid   Mexico City
Milan   New Delhi   San Juan
Seoul   Singapore   Sydney   Toronto

Sponsoring Editor
Roger Stewart

Editing Supervisor
Stephen M. Smith

Production Supervisor
Pamela A. Pelton

Acquisitions Coordinator
Amy Stonebraker

Project Manager
Patricia Wallenburg, TypeWriting

Copy Editor
James Madru

Proofreader
Claire Splan

Indexer
Jack Lewis

Art Director, Cover
Jeff Weeks

Composition
TypeWriting

This book is dedicated to the audio mastering engineers who have helped teach others, such as Bob Katz, who skillfully authored *Mastering Audio: The Art and the Science*; Dave Collins, who has always shared his unparalleled knowledge with others; and Brad Blackwood, mastering engineer at Euphonic Masters, Memphis, TN.

About the Author

Gebre Waddell is a noted mastering engineer and owner of Stonebridge Mastering. His 10-year career includes work with artists such as Public Enemy, Ministry, George Clinton, and Rick Ross. He is a member of the Grammy Recording Academy and the Audio Engineering Society. Mr. Waddell has created many resources, including *The Digital Publishing Standard for Audio Recordings* and a popular spectrographic frequency chart titled "The Frequency Domain." Most recently, he has been working with DSP algorithm design and VST programming.

Contents

Introduction   xix
Acknowledgments   xxi

1 Overview of Mastering and the Typical Session   1
Overview of Mastering   1
    Goals of Mastering   1
    Role of the Mastering Engineer   2
    Choose an Engineer Based on Genre?   2
    Who Needs Mastering?   3
    Don't Master Your Own Mixes   3
    Balance of Benefit and Sacrifice   3
    Creativity in Mastering   3
    Subtlety   3
    Mastering Is an Art   3
Typical Mastering Session   4
    The Engineer   4
    Receiving   4
    Attended/Unattended Sessions   4
    Acoustics and Monitoring   4
    Passes   5
    Processing   5
    Final Output   5

2 Do-It-Yourself Guide for Basic Mastering   7
The Most Basic CD Mastering   7
    Issues with Such a Basic Approach   7
    MP3 Encoding Without Other Processing   8
Selecting a Digital Audio Workstation   8
Basic Mastering Processing Chain   8
Processing Packages   8
Basic Equalizer   8
Basic Compression   9

Basic Limiting 9
Compare the Processed Version with the Original 10
Sample-Rate Conversion 11
Dithering 11
Export to CD 12
Exporting for Internet Release 12
This Portion Is Only a Basic Guide 12

3 Preparing a Mix for Professional Mastering **13**
Maximum Peaks at –3 dB 13
Mixing into a Limiter 14
Requesting the Removal of Mixbus Processing 14
Multiple Mono versus Stereo Interleaved 14
Selecting Stem Mastering 14
Allowing Time for Mastering 15
Sample Rate 15
Mastering to Tape 16
Bass and "Air" 16
Mix Problems 17

4 Accepting Mixes, Workflow, and Client Interfacing **19**
Receiving/Importing Digital Recordings
   (Bit Rate, Sample Rate, and Levels) 19
Data Integrity: CRC and Checksums 20
Requesting Information from the Client 20
Receiving/Importing Analog Recordings 20
Attended versus Unattended Sessions 20
Organizing Our Work 21
      Saving/Copying/Pasting Processing Configurations 21
      Importance of Timeliness 21
Each Recording Is Processed Individually 22
File Compression 22
File-Transfer Options 22
Focus on Customer Service 22
Time Frame for Revisions in Mastering 22

5 Mastering Gear **25**
Gear Lust 25
Equipment Demos 25
Stepped versus Continuously Variable Controls 26
Detented Potentiometers versus Rotary Switches 26
Analog Components 26
Software Plug-ins versus Analog Processing 26
Mastering Software Bundles 27

Native Processing 27
Retailers, Auctions, and Classified Ads 27
Acoustic Treatment 27
Monitoring Control Systems 28
Monitors (Speakers) 28
    Ideal Monitors 28
    Passive versus Active Monitors 28
    Full-Range versus Midfield versus Near-Field Monitors ... 29
    Upgrading to 5.1 29
    Crossovers 30
    General Placement 30
    Monitor Positioning 30
Subwoofers 31
Monitor Amplifiers 31
A/D and D/A Conversion 32
Audio Interfaces 34
Wordclocks/Master Clocks/Distribution Amps 35
Analog Equalizers 36
Plug-in Equalizers 37
Dynamic Equalizers 38
Compression/Expansion 38
    Compression 38
    Expansion 38
    Parallel Compression 39
    Side-Chain 39
    Selecting a Compressor 39
    Analog Compressors 39
    Plug-in Compressors 41
    Multiband Compressors 41
Maximizers/Limiting 42
Stereo/Mid-Side Processors 42
    What Is Mid-Side? 42
    Stereo Processors 43
    Analog M/S Converters 43
    Digital Stereo Processing 43
De-Essers 44
    Analog De-essers 44
    Digital De-essers 44
Restoration and Noise Reduction 44
Harmonic Enhancement and Saturation 45
    Analog 45
    Digital 45
Routers/Patchbays 45
Consoles 46

Headphones 47
Acoustic Environment Simulation for Headphones 47
Digital Audio Workstation Software 47
Playback Sources 48
DDP Software 48
Sample-Rate Converters 49
    Software 49
    Hardware 49
5.1 Mastering 49
Meters 50
    Analog 50
    Digital 50
Mastering DSPs 51
Combination Processors 51
Metadata-Embedding Software 51
Forensic Audio Software 52
Connections and Cables 52
    Analog XLR 52
    AES/EBU 53
    S/PDIF 53
    1/4th, 1/8th, and Minijacks 53
    Multi-Pin Connectors 53
    RCA 53
    BNC Wordclock 54
    Bantam/TT 54
    ADAT Lightpipe 54
    USB 54
    FireWire 54
    Thunderbolt 54
    PCI/PCIe 55

6   Mastering Acoustics and Monitoring 57
Hiring a Professional Acoustician 57
Room Modes 57
Absorption 58
Absorption Coefficients 58
Diffusers 58
Front-Wall Treatments 58
Treating First Reflection Points 59
Clouds and Ceilings 59
Narrowing the Front of the Room 59
Minimizing Noise 59
Rear Diffusers 60
Early Reflections 60

Listen to Tones with an SPL Meter 60
Parallel Surfaces 60
Console/Desks/Surfaces 60
Symmetry 61
Decoupling Monitors 61
Bass Traps 61
Minimal Objects in the Room 61
Ideal Room Dimensions 61
Angled Walls Behind Monitors 62
Monitor Visibility 62
Correcting Acoustic Problems with Equalization 62
Speaker Placement 63
Speaker Decoupling 64
Integrating Subwoofers 64
Subwoofer Placement 64
Subwoofer Calibration 64
    Method A 65
    Method B 65
    Method C 65
    Method D 65
    Subwoofer Phase 65
    Excerpt from the Dolby 5.1-Channel Production Guidelines
        (the LFE Channel Is the Subwoofer) 65
RT60/RT30/RT20 66
Listen to Translations 66
Acoustic Methods/Techniques/Design Concepts 66
Selecting Monitors 67
High-Quality Amplifiers 67
Biwire 67
K-Meter Explained 69
K-20, K-14, and K-12 69
Stepped Monitor Gain Control 70
K-System Criticisms 70
Consistency as a Result of Reference Levels 70
Monitoring at a Variety of Loudness Levels 70
Ear Sensitivity 71
Alternative Reference-Level Selection 71
Workplace Safety 72
Headphones 72

7   **Mastering Practices: Techniques, Problems, and Approaches** .. **73**
Technique: Listening 74
Technique: Working with Intent and Vision of the Result 74
Technique: Only Making True Improvements 74

Approach: Destructive versus Nondestructive Processing 76
Technique: Working with Reference Recordings 77
Technique: Minimizing the Delay Between Comparisons 77
Technique: Avoiding Ear Fatigue 78
Technique: Processing Sections of a Song Separately 78
Technique: Minimizing Processing 79
Technique: Making Client-Requested Changes Properly 79
Technique: Processing Based on First Impressions 79
Technique: Turning Things Off/Listening in Bypass Mode 79
Technique: Relationships with Mixing Engineers and Producers 80
Technique: Concurrent Processing 80
Technique: Processing in Stages 80
Technique: Stem Mastering 81
Technique: Reverb Processing 82
Technique: Mastering Equalization 82
    Parametric Equalizer Controls 83
    Analog Equalizers 83
    Digital Equalizers: Minimum Phase 83
    Digital Equalizers: Linear Phase 83
Technique: Using a Graphic Equalizer 84
Technique: Basic Frequency Balancing Using an Equalizer 84
    Subsonic (~0 to ~25 Hz) 84
    Bass (~25 to ~120 Hz) 84
    Lower Midrange (~120 to ~350 to 400 Hz) 85
    Midrange (~350 to ~2,000 Hz/2 kHz) 85
    Upper Midrange (~2 to ~8 kHz) 85
    Highs (~8 to ~12kHz) 86
    "Air" (~12 kHz to the Limit of Hearing) 86
    Out-of-Band Noise 86
    Filters, Shelves, Bells, and Q Values 87
    Final Word 88
Technique: Order of Frequency Adjustment 88
Technique: Substractive Equalization 88
Technique: Using Less Common Equalizer Filters 89
    Baxandall Shelves 89
    Gerzon Shelves 89
    Niveau/Tilt Filter 89
Technique: Frequency Roll-Off on Both Ends 90
Technique: Extra Equalizer After Compression 90
Technique: Monitoring the Middle and Side Channels 90
Technique: Mid-Side Processing 91
Technique: Checking Mono Compatibility 91
Technique: Using Unique Mid-Side Processors 92
    DDMF Metaplug-in/Mid-Side Plug-in 92

iZotope Ozone 92
Brainworx 92
Mathew Lane's DrMS 92
Technique: Understanding Distortion/Coloration/Saturation .. 92
Technique: Using Digital Emulations of Classic Gear 93
Technique: Running Through Twice 93
Technique: Adding Distortion to the Side Channel 94
Technique: Using the Same Character Processors on All Songs .. 94
Technique: Using Dither 94
Dithering in the Visual Realm 94
Audition Different Types 95
Impact of Dither on the Sound 95
Dithering Should Be Kept to a Minimum 95
Dithering for Digital Processing 95
Problem: Jitter 95
Problem: DC Offset/Asymmetrical Waveforms 96
Technique: Bass Enhancement 96
Problem: Bad Mixes 97
Problem: Hum in the Analog Signal Chain 97
Problem: Sibilance 98
Technique: Raising Levels Before Analog Processing 98
Technique: Adding Noise in the "Air" Band for Brilliance 99
Problem: Limiting Distortion 99
Problem: Intersample Peaks 99
Technique: Using Sample-Rate Conversion 100
Problem: Using Unbalanced-to-Balanced Connections 100
Problem: Unbalanced Left and Right Channels 100
Problem: Unbalanced Bass in the Mix 100
Technique: Upsampling 101
Technique: Digital Limiting 102
Problem: Lack of Vocal Clarity 102
Problem: Matching Equalizers 102
Problem: Harsh/"Digital" Sound 103
Addressing Resonances 103
Raising the Lower Midrange or Bass 103
Rolling Off the Highs 103
Using a High Shelf 103
Dynamic Equalization on Upper Midrange/Refinement ... 103
Using a De-Esser 104
Analog Processing with Tubes or Transformers 104
Warming with a Compressor 104
Problem: Muddiness 104
Problem: Part of the Frequency Spectrum Is Out
of Balance During Loud Passages 105

Problem: One Part of the Frequency Spectrum Is Too Dynamic . . 105
Problem: A Broad Part of the Frequency Spectrum Is Too High . . 105
Problem: Less Than Full Sounding 105
Problem: Fast/Medium Transient Sounds Stick Out Too Much . . 106
Technique: Manually Reducing Peaks 106
Technique: Reducing Level Before a Part Change 106
Technique: Analog-and-Digital Gain Staging 106
    16-, 24-, and 32-Bit Formats 106
    Digital Gain Staging 107
    Analog Gain Staging 107
Technique: Mastering with a Focus on the Vocal 107
Technique: Working with a Vocal-Up Mix 108
Problem: Lacking Depth, Needs a Three-Dimensional Sound . . 108
Problem: Recordings Sound Different in the Car 109
Technique: Mixing Down to Tape/Mastering with Tape 109

8 **Shaping Dynamics** **111**
Types of Dynamics Processing 111
    Compression/Downward Compression 112
    Expansion/Upward Expansion 112
    Parallel Compression 112
    Side-Chain Compression 112
    Multiband Compression 113
    Limiting 113
Compression Settings and Meters 113
Setting Attack/Release/Threshold 113
Macrodynamics/Microdynamics 114
RMS-Sensing Compressors 114
Peak-Sensing Compressors 114
Compressor Response 115
Character versus Transparent Compression 115
Mixbus Compressors 115
Punchy Compression 115
Serial Compression 116
Volume Automation for Macrodynamic and
    Microdynamic Adjustments 116
Gain-Reduction Meters 116
Compressor Input Levels 116
Compressor Output/Makeup Gain 117
Linked versus Unlinked Compression 117
Expansion/Expansion Before Compression 117
Analog Compressors 117
    Voltage-Controlled Amplifier (VCA) 117

Opto/ELOP 118
Variable Mu 118
Field-Effect Transistor (FET) 118
Pulse-Width Modulation (PWM) 118

9   **Achieving Loudness** **119**
Apparent/Perceived Loudness 119
Beyond Ideal Loudness, There Is Quality Loss 120
When Did the Loudness War Begin? 120
Why Does the Loudness War Exist? 120
Future of the Loudness War 120
Loudness Potential of a Recording 121
Digital Clipping 121
Using Clipping 121
Compression Before Limiting 121
Digital Limiting 121
Operating a Limiter 122
Sensitivity of the Ear 122
Clipping a High-Quality A/D Converter 122
Serial Limiting 123
Digital Limiter Ceiling of –0.3 dBFS 123
Maximizers/Multiband Limiters/Inflators 123
Ideal Loudness 124
Processing While Focusing on the Loudest Passages 124
True Peak Ceilings 124
Broadcast Loudness Standards 124
ITU-R BS.1770 125
EBU R128 125
ATSC A/85 125
CALM Act 125

10 **Fades, Sequencing, and Spacing** **127**
Spacing Between Songs 127
Performing Fades/Cross-Fades 128
Fades During Mixing Stage 128
Fades During Mastering Stage 128
Performing Fades During Mastering Has
Its Own Set of Benefits 128
Sequencing an Album 129
Noise Reduction 129
Denoise First 129
Noise at Beginnings and Endings 129
Learn Feature 129

Noise/Hiss/Clicks/Pops 129
Spectral Editing 130
Careful Application 130

11   **Visualizations/Metering** **131**
FFT/Fourier 131
Using Spectral Analyzers 132
Bit Meters 132
Correlation Meters 133
Vectorscope 133
Reconstruction Meters 134
Meter Action/Speed 134

12   **Preparing the Final Output** **135**
Providing Client a Preview for Approval 135
Making Revisions 136
Quality Control 136
Red Book/Rainbow Books 136
Red Book CD Specifications 137
Setting Track Markers 137
CD Pause Length 137
Track Offsets 138
International Standard Recording Codes 138
MCN/UPC/EAN Codes 138
CD-Text 139
Premaster CD 140
Writing Speed 140
Disk-at-Once/Track-at-Once 140
Error Checking with Plextor/Plextools 140
CD Error Levels 141
DDP File 142
DDP versus Premaster CD 142
Zipping/Archiving the DDP 142
BIN/CUE (An Alternative to DDP) 142
Other DDP Alternatives 143
Drawbacks to DDP Alternatives 143
Mastering for Vinyl 143
Shipping to the Client 143
PQ Sheets 144
Shipping to the Replicator/Duplicator 144
Checksum/MD5 144
Nonlossy: WAV, AIFF, FLAC, Etc. 145
Nonlossy Metadata 145

Lossy: MP3 146
DVD-V, DVD-A 146
SACD 146
Blu-ray Audio 146
Mastered for iTunes 147
Enhanced CD 147
Ringtones 148
5.1 Audio 148

13   What Happens After Mastering? **149**
Storage and Returns 149
Database Submission 149
All-Music Credits 150
Replication versus Duplication 150
Eclipse Systems 150
Disc Life 151
Radio and Broadcast Processing 151
Car Stereo 151
CD Refractive Index 151
Clubs 152
Growing Home Theater 5.1 Systems 152
Internet/Streaming/Format Conversions 152
Smart Phone/MP3 Player/Computer Playback 152

14   DAW/Computer Optimization and Interfacing **153**
Do Not Let a DAW Dictate the Workflow 153
Save Early, Save Often 154
Keyboard Shortcuts/Hotkeys 154
DAW Functions 154
Wacom Tablets/Trackballs/Mice 154
RAID and Online Backup 154
Solid-State Drives (SSDs) versus Hard-Disk Drives (HDDs) 155
Ending Nonessential Processing and Services (Windows) 155
User Accounts 155
Faulty Drivers 156
Plug-in/Interface/DAW Conflicts 157
Other Typical Problems/Regular Maintenance 157
DPC Latency Checker 157
Computer Hardware Problems 157
Operating System Tweaks 158
Disable Onboard Sound in BIOS 158
Disable Internet and Antivirus 158
Second Hard Drive for Audio 158

FireWire   158
Startup   159
System Registry   159
Latency/Buffers   159
Driver Systems   159
Spyware/Badware/Viruses/Malware   159
Be Serious About Technical Problems   160

15 Starting a Mastering Studio as a Business **161**
Challenges Faced by New Mastering Studios   161
Entrenched Competitors   161
Do-It-Yourself (DIY) Mastering   162
Customer-Service Challenges   162
False Advertising and Scams   162
Album Sales at All-Time Low   162
Nonscalable   163
Low Growth Rate   163
High Startup Cost   163
No Points   163
Limited Scope   163
Legal Disclaimers/Limited Liability   163
Benefits of Starting a Mastering Studio   164
Talent in Practice   164
Marketing Without Limitation   164
Surge of Independent Artists   164
Shorter Projects   164
Less Band Politics   165
Local Attended Sessions   165
Being a Sole Proprietor   165
Rising to the Challenge   165
Seek an Internship   165
Take a Course   165
Collect Resources   165
Learn Aggressively   166
Learn About the Legends   166
Add Mastering Services to an Existing Studio   166
Leverage Existing Opportunities   166
Reach New Markets   166
Visit the National Association of Music Merchants (NAMM),
    Join the Audio Engineering Society (AES) and
    Grammy Recording Academy   166
Working for Notable Acts   167
Insure Your Studio   167

Hiring Interns   167
Other Opportunities   167
    Opportunities for Audio Forensics   167
    Consider Related Fields   167

16   Contributions **169**
On Analog Multiband Compression and Audio Gear   169
    Robin Schmidt, Mastering Engineer
    Owner of 24-96 Mastering, Karlsruhe, Germany

The Case for Full-Range Monitoring   172
    Scott Hull, Senior Mastering Engineer
    Owner of Masterdisk Mastering Studios, New York, NY

    What Is That?   173
    But What About the Consumer?   174
    What Are My Goals in Monitoring?   174
Connection and Calibration of an Analog Mastering Chain   175
    Jaakko Viitalähde, Owner and Mastering Engineer
    Virtalähde Mastering, Kuhmoinen, Finland

    Prologue   175
    Building Up a Chain   175
    Patching Methods   177
    Installation, Calibration, and Operating Levels   178
    Closing Words   179
Distortions and Coloring   180
    Dave Hill, Audio Engineer and Owner of Dave Hill Designs
    Equipment Designer for Crane Song, Ltd.

    Introduction   180
    Distortions as Color   181
    Nonlinear   182
    Time Domain   183
    Coloring   183
Mid-Side Processing   184
    Brad Blackwood, Mastering Engineer and Owner
    Euphonic Masters, Memphis, TN

Digital Filtering   185
    Pieter Stenekes, Founder and Owner
    Sonoris Audio Engineering, Friesland, Netherlands

    What Is a Filter?   185
    Digital Signal Processing   185
    Filter Types   187
    Design and Implementation   190
    Conclusion   191

Optimizing Audio for Radio 191
  Cornelius Gould, Audio Processing Developer
  Omnia Audio, Cleveland, OH

Premastering for Vinyl Cutting 194
  Jeff Powell, Direct Vinyl Transfer Engineer
  Engineer for Stevie Ray Vaughn, Bob Dylan, and Many More Artists
    What If the Songs Will Not Fit? 195
    Processing for Master Lacquer 196
    Cutting the Master Lacquer 197
  ASIO versus WDM 198
  David A. Hoatson, Cofounder and Chief Software Engineer
  Lynx Studio Technology, Inc.
    History and Implementation 198
    Sonic Differences 199
    Zero-Latency Monitoring 200

A   Decibel Units of Measure 201

B   Mastering Resources and References 203
    Online Resources 203
    Offline Resources 204
       Print 204
    Video 205
    Classes 206

    Index 207

Introduction

This book is a great place to start for the beginning mastering engineer, a valuable set of insights for the musician, and a helpful resource for the advanced engineer. It is a comprehensive resource about audio mastering focused on practical ideas and techniques.

Professionals in any discipline will tell you that it is really impossible to capture the subtleties of that profession in a book and that practice is the key. This is certainly true for audio mastering. Nevertheless, for the dedicated reader, this book will provide insights into the art and science of audio mastering like no other guide.

This book is written with two main goals in mind. First, I wish to give readers a rich, all-inclusive description of the concepts and equipment related to mastering. Ideally, professionals should be aware of most everything involved with their field—even things they do not choose for their own work. The second goal is to put the information into perspective. For this, there are thoughtful explanations and insightful contributions from respected professionals, imparting mastering wisdom on key subjects.

Presenting the concepts of audio mastering this way allows readers to gain a deep understanding—an understanding that allows for an emotional, instinctual approach to professional audio mastering. As the concepts are internalized and experience takes its course, an engineer's style develops.

Some readers may be looking for mastering "secrets," and while there is actually no big secret, this book attempts to leave no stone unturned. The fact is that mastering is about the sum of many small things that together make for an extraordinary sound.

I am offering to readers two free VST plug-ins, both of which are referenced in this book. The Precision Post-Analog Controller is designed for making precise gain adjustments and the Audio Calibration Tool includes several test tones, noise types, and other basic tools used for calibration. These plug-ins are available at: http://www.stonebridgemastering.com/plugins.

Gebre Waddell

Acknowledgments

I would like to thank my wife, Dawn Waddell, and my family, Mary Ann Waddell, James Waddell, and Kisa Seymore.

I would like to give a very special thanks to William Small for helping me in so many ways and Khari Wynn for his significant support.

I also would like to thank individuals who have helped me in some way while I have been on this path, including Roger Stewart, Angelo Earl, Tito Jackson, Chuck D, Nil Jones, Burgess Macneal, Alex Alexzander, Collin Jordan from Boiler Room Mastering, Jason Ward and Bob Weston from Chicago Mastering Service, Jeff Powell, Susan Marshall, Sean Rickman, Hutch Hutchison, Justin Short, John Scrip, Larry Nix, Kevin Nix, Beau Whitlock, Erica Paschall, Chris Riggs, Jesse Spiceland, John-Paul Laborde, Erica McDaniels, Jeff Peck, Jason Peck, Austin Harris, Don Roby, Robert Allen Parker, Crafton Barnes, Eric Gales, Danielle Gold, Boo Mitchell, Robert Chatham, Jahleel Eli, Alfred Waddell, Scott Leader, Tom Sixbey, Ari Morris, Steve Munns, Anthony Crawford, Tom Menrath, Ping D. Rose, Clifton Harviel, John Miller, Marcela Pinilla, Lilith Andrade, Langston Wynn, Marc Thériault, Marc-Olivier Bouchard, and Anthony Basurto.

Thanks goes to Dave Collins, who reviewed this book while making notes, edits, and suggestions. Thanks to Huntley Miller for technical proofreading. Also, thanks to Sophocles Orfanidis for helping in the final stages of the book.

CHAPTER 1

Overview of Mastering and the Typical Session

Overview of Mastering

It takes several stages of production to create the recordings we know and love. These include songwriting, recording/tracking, mixing, and mastering. *Mastering*, the last step, is the finalization of audio recordings and the preparation for their final medium. Where mixing involves working with individual tracks, mastering usually involves working with stereo mixes, which are just two channels, left and right. What exactly happens during mastering? There are three primary processes— equalization, compression, and limiting. Then, usually compact discs [CDs, or Disc Description Protocol (DDP) files used to manufacture CDs], MPEG-1 or MPEG-2 Audio Layer III files (more commonly known as MP3s), or Waveform Audio File Format (WAVE or, more commonly, WAV owing to its file extension) files are sent to the client. This chapter gives an overview to help you gain a good general sense of mastering.

Goals of Mastering

Once the mastering session has started, we must do what sounds best. Many mastering engineers would agree that one of the highest goals of mastering is to ensure the best translation. *Translation* is how a recording sounds on a variety of playback systems. We also must be sure to maximize the emotion of the recording at hand and process with musicality. Figure 1-1 shows a few of the playback systems that someone may use to listen to the final mastered recording.

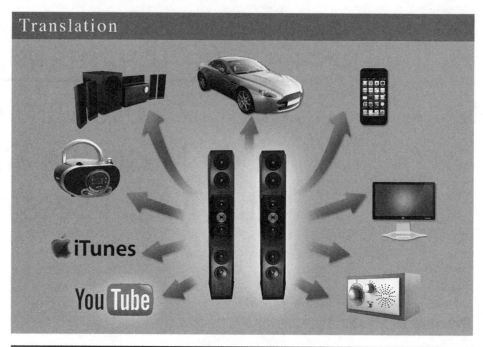

FIGURE 1-1　One of the highest goals in mastering is to ensure the best *translation*, which is how a recording sounds on a variety of playback systems.

Role of the Mastering Engineer

Over the years, there have been many metaphors that describe mastering. One of the best compares the mastering engineer to a team owner and the mixing engineer to a team coach. The coach knows all the personalities and the strengths and weaknesses of the players. The mastering engineer is more like the team owner. He or she doesn't know every detail, and he or she shouldn't. Instead, the mastering engineer is looking out for the overall success of the team. The differences in perception and the variety of roles are important for creating something great.

Choose an Engineer Based on Genre?

Some mastering engineers are known for their techniques; others are known for certain genres, such as dance music, orchestral, or pop mastering. However, even the engineers known for working with a certain genre are usually just as skilled at working with all styles of music. Mastering is far less genre-specific than production or mixing.

Who Needs Mastering?

Recordings are made at so many different levels of quality. Some are created by seasoned recording engineers in amazing studios and some in small home studios. Mastering holds benefits for novice and professional productions alike. While novice productions stand to gain the most from professional mastering, almost no professional would ever go without it.

Don't Master Your Own Mixes

Mastering is best performed by those who make it their life's work. The world's greatest mixing engineers, almost without exception, do not perform mastering on their own mixes. They trust their mixes only to professionals who are dedicated to mastering. Mastering is an area of specialty, and the skills simply take time to develop.

Balance of Benefit and Sacrifice

Because every process affects the entire mix, mastering virtually always involves a balance between benefit and sacrifice. High-quality mastering requires a sense of priority and seasoned judgment.

Creativity in Mastering

Creativity varies in mastering, depending on the source recording and the engineer's tastes. Generally, professional mixing engineers tend to want more transparent processing (less creativity by the mastering engineer), whereas amateurs often want mastering to transform their sound.

Subtlety

When someone mentions mastering, subtlety comes to mind. Fine adjustments are often all that is needed when applying processing over an entire mix. As they say, a little goes a long way.

Mastering Is an Art

There is no standardized method of mastering. Professional mastering engineers have their own unique approaches and develop a personal sense of how to achieve their vision. Great mastering engineers have great taste for sound quality. Mastering engineers develop their techniques and equipment chains for years to create a sound that satisfies their tastes.

Typical Mastering Session

After quite a bit of hard work, the album's mixes are finished up and ready for mastering. Now it's time to choose a mastering studio. The mastering session will soon begin, and the album will be ready to release to the public. Let's talk about what to expect as the mastering session unfolds.

The Engineer

When you first meet a mastering engineer, how do you know if he or she is good? What makes one better than another? Well, the most important factors in mastering are the skill, ears, talent, and experience of the mastering engineer. These all-important factors can be a difficult thing to gauge when choosing a service. Because of this, most people evaluate a mastering engineer by listening to their previous work.

Receiving

Professional mastering involves a client (usually an artist, producer, or mixing engineer) giving his or her recordings to a mastering engineer in person, by shipping, or via the Internet. These recordings are almost always digital files or analog tape. Usually, an order form is submitted, or the details are otherwise provided.

Attended/Unattended Sessions

Among top engineers, opinions vary about attended and unattended sessions. There is a great benefit to be had from sessions without the client being present. The mastering engineer is most familiar with his or her monitoring environment and can focus entirely on the task at hand. With attended sessions, clients may insist on changes that may be regretted later when they listen outside the studio. On the other hand, attended sessions allow for more personal communication between client and engineer and can foster the business relationship. Also, some mastering engineers view their attended sessions as a bit of a performance that clients enjoy.

Acoustics and Monitoring

So the mixes are playing and the mastering adjustments are being made. Both the speakers and the room acoustics are affecting what you hear, which of course, will affect the adjustments you make. *Acoustics* are the impact of a room on a sound being made inside it. In professional audio, speakers are called *monitors* because they allow you to monitor the sound.

Mastering is performed best in a well-designed acoustic environment with an accurate monitoring system. The acoustics should have as little impact as possible on the frequency balance. This means that sounds all across the frequency range will have a similar loudness. Accurate monitoring and acoustics allow the mastering engineer's adjustments to translate well to the widest variety of playback systems.

Passes

Mastering engineers must listen to a recording as they make adjustments. Each time a mastering engineer listens to a song from start to finish while making adjustments, it is called a *pass*. A few mastering engineers price their sessions according to the number of passes they perform.

Processing

A mastering engineer uses three main tools for processing—equalization, compression, and limiting. *Equalization* is perhaps the most important. It is a process that affects the frequency response of a recording. For example, with equalization, the bass or treble can be raised.

Compression and limiting are dynamics processing. Basically, *dynamics* in the audio world describes how much loudness changes over a period of time. This may be a very short period of time, such as milliseconds, or longer, such as a second or even a minute.

Final Output

Okay, so the mastering is all done. Now what? The most common final media are a *Red Book* audio CD, *Disc Description Protocol* (DDP), MP3, and WAV.

Red Book audio CDs are made in such a way that they are ready for duplication/replication. These CDs are also called *premaster CDs* (PMCDs).

A DDP file is a computer file that contains all the information of a CD, including CD-Text. The DDP file can be sent electronically to a duplicator/replicator without the need for shipping.

Sometimes clients request other formats, such as MP3 files encoded from the 24-bit master or 24-bit WAV files specially prepared for Apple's Mastered for iTunes program.

Often *PQ sheets* are provided, which are lists of album and track information. *Error sheets* are also common, which certify that CD error levels are consistent with the *Red Book* standard or the more stringent requirements of duplicators/replicators.

CHAPTER 2

Do-It-Yourself Guide
for Basic Mastering

This chapter covers the things you need to know to get started with mastering right away. Sometimes a fast start is the best way to begin with something new. Also, mastering seems to appeal to those who are naturally hands-on. Even for readers who are experienced with mastering, this chapter may help to frame the concepts of the rest of the book.

The Most Basic CD Mastering

Let's start with the most basic approach for creating a CD. You would collect all the mixes, arrange them in the right sequence, and burn them on a CD with one of the many applications available for the task. That's it, you can play the CD.

Issues with Such a Basic Approach

With the previous approach, when you listen to the CD, the tonality may be far different from that of professional recordings, and to some people, tonality is everything. If the mixes were 24-bit and considering that CDs are 16-bit, bit truncation would have occurred. Also, the loudness may be uneven and lower than that of commercial releases. Professional duplicators/replicators may refuse to work with the CD for a number of reasons. Basically, it might not sound professional and is not yet ready for duplication/replication. These are some of the main issues with which mastering is concerned.

MP3 Encoding Without Other Processing

Instead of burning a CD, you may instead make an MP3 file from the 24-bit mix. This avoids the problem with truncated bits, but still the tonality and loudness issues remain.

Selecting a Digital Audio Workstation

A digital audio workstation (DAW) is a computer program that helps users work with audio. Some DAWs have been specifically designed with mastering in mind. Magix Samplitude and Sequoia, both of which run on the PC only, are great choices for mastering engineers at all levels. A complete list of mastering DAWs is included in Chapter 5.

Basic Mastering Processing Chain

Digital audio workstations (DAWs) usually include equalizers, compressors, limiters, and dithering. Also, DAWs have a way to arrange the processing in any order you wish. For mastering, a good starting point is a sequence of equalization, compression, limiting, and then dithering. The processing sequence is very important in mastering.

Processing Packages

There are several digital mastering plug-in packages that contain a suite of audio mastering processors. These include iZotope Ozone, T-Racks, Plugin Alliance Big-4 Bundle, and Waves Masters Bundle. These plug-ins work with a variety of DAWs. Each is shown in Figure 2-1, and any one is perfect for getting started.

Basic Equalizer

The *equalizer* is the main tool for adjusting tone, which is a major part of mastering. Perfecting the art of professional equalization is something that takes considerable time. For quick-start purposes, I will list a few basic techniques.

Plug-in equalizers usually sound best if they have a minimum phase mode and it is selected (as with the DMG EQuality plug-in). If necessary, try using parametric equalization with a medium Q (a value of 1-2, but perhaps as high as 4) to reduce any resonant frequencies in the lower midrange (120–400 Hz). Resonant frequencies are actually quite rare, but if present, they sound like a

Figure 2-1   iZotope Ozone, T-Racks, Plugin Alliance Big-4 Bundle, and Waves Masters Bundle are popular digital mastering plug-in bundles.

ringing and may affect the entire spectrum. You may boost to find them and then cut to reduce or remove them.

Harshness is very common, unlike resonant frequencies. Use a parametric equalization bell, at 3.15 kHz, with a Q value of 3.5 to reduce harshness.

A bell with a broad width (lower Q value) may be used to affect large areas of the spectrum.

Use shelves to adjust entire ranges, such as the entire bass range (120 Hz and below) or the entire high-frequency range, if necessary. No adjustment should be made unnecessarily. These techniques are shown in Figure 2-2.

Basic Compression

When just starting out, barely use any compression, and consider using none at all. Compression is basically an automatic loudness control. It's much less of a factor in professional mastering than most people think.

Basic Limiting

Limiters are processors with two main controls: a gain control and a ceiling. The *gain control* is used to adjust the loudness of the recording, and the *ceiling* is the maximum level that can occur. The ceiling has much less of an impact on the

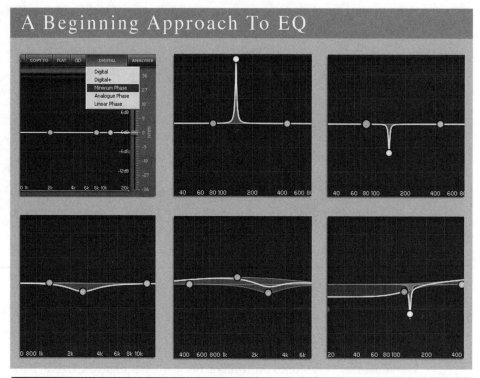

FIGURE 2-2 Minimum phase option; boost to find a resonance; cutting a resonance; 3 kHz cut; broad equalization; using shelves to adjust ranges.

perceived loudness than the gain control. This is so because it is the *average* level that gives the perception of loudness much more than the *maximum* level.

Okay, so how do you set the limiter? The limiter's gain control should be set for each recording in such a way that when you listen, each recording of the album or extended play (EP) has the same or similar loudness. You might compare your limiting to a recording in a similar genre. You also might just set the limiting where loudness is maximized but without too much of the negative side effects that can come from limiting. Also, it is very typical that the output ceiling is set to –0.3 dB.

Compare the Processed Version with the Original

When comparing the processed and original recordings, each must be compared at the same loudness. This task is a little more complicated when using analog equipment and will be covered in Chapter 7 in more detail. When mastering with digital plug-ins, comparisons are a bit easier. A processor with a gain control can be added at the very end of the sequence of processors. Gain controls allow the loudness of a recording to be altered. Basically, you'll be using this gain control at the end of the chain so that when all plug-ins are bypassed, the loudness is the same as when they are all active. Be honest about making the loudness the exact same; otherwise, the louder version may always seem better, even when it is not. Remember to bypass or delete the processor used for loudness matching when exporting. An example of such a processor is shown in Figure 2-3.

Sample-Rate Conversion

A sound is represented in the digital world with many points called *samples*. The *sample rate* is the number of digital samples in time that make up an audio waveform. DAWs usually display the sample rate, so it's easy to find. If the audio sample rate of the recording at hand is not 44,100 (often abbreviated as 44.1k), then it must be converted to create a CD. Conversion to a sample rate of 44.1k is usually preferred to be done in sequence just before dithering.

Dithering

While the sample rate is the number of digital samples, the *bit rate* is the range of the samples. *Dithering* is noise added in with audio to help reduce the effects of converting from a higher bit rate to a lower one. Dithering is performed as the last step before converting to a lower bit rate. This is important because mastering

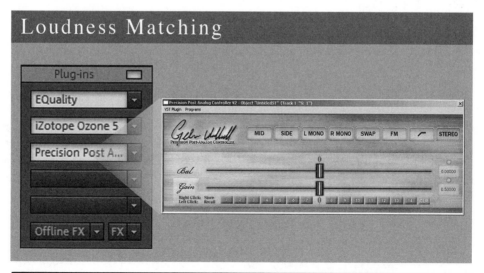

FIGURE 2-3 Using a tool such as Gebre Waddell's Precision Post-Analog Controller to carefully match loudness is essential for making comparisons.

processing is typically performed at 32 or 24 bits, and the final output is 16 bits. Many limiters have a dithering feature, as well as most DAWs.

Export to CD

It's easy to create a *Red Book* audio CD; you just need the right software. Just export a WAV file at 16 bits, and burn it with CD-burning software or use the burning feature of your DAW. Use burning software that allows the creation of a *Red Book* master and that allows you to enter CD-Text details.

Exporting for Internet Release

If a recording is intended for an Internet release, then usually an MP3 file will be needed. MP3s are best created from your 24- or 32-bit sources instead of from the 16-bit CD files. After the mastering has been performed, most DAWs have an MP3 export option that can be used to easily create MP3 files.

This Portion Is Only a Basic Guide

Please keep in mind that this portion of the book is a basic guide. It helps in getting started and helps frame the topics of the book.

CHAPTER 3

Preparing a Mix for Professional Mastering

Okay, so the mixes are finished and seem ready for mastering. There may be a few things you can do to make sure that they reach their full potential. In this chapter you will find out about some of the preparation professionals do to get the best out of the mastering process.

Maximum Peaks at –3 dB

Providing mixes for mastering that peak at a maximum of –3 dBFS (decibels relative to full scale) has become somewhat of a standard practice. To accomplish this, the mixing engineer can adjust their master fader during mixing until the highest peak shows on a digital peak meter at –3 dBFS. For the most part, this is a preventative measure. It provides a comfortable margin to ensure that there is no clipping. If 32-bit float resolution is used and is provided to the mastering studio, overloads are not possible. Even so, the –3-dBFS level is still usually observed.

NOTE *FS (full scale) is added after dB to indicate a digital measurement of decibels because there are many types of decibel units of measure. While working in a DAW, virtually every dB abbreviation you see is actually dBFS. Another commonly discussed decibel type is dBSPL, which indicates a measurement of the sound pressure level using a microphone in the real world. For a comprehensive list, see Appendix A.*

Mixing into a Limiter

Some mixing engineers mix with a transparent limiter on the master bus (such as Voxengo Elephant, Slate Digital's FG-X, or Ozone's IRC III). The master bus (also called the *mixbus*) is a channel in the DAW through which the entire mix is routed before it is heard or exported to a file. Adding a limiter to the master bus gives a sense of how the mix may respond to mastering. Master bus limiters also can be used to make more accurate reference comparisons between mixes and commercially released masters. The limiter is removed or bypassed before submitting the mix for mastering or somehow an alternate mix is submitted for mastering without the limiter.

Requesting the Removal of Mixbus Processing

Whereas some mastering studios request that all mixbus equalizer/compression processing be removed before mastering, it is not necessarily a good standard practice. The reasoning behind the idea is that novice engineers can harm the potential of their recordings by processing in ways that cannot be undone. However, experienced engineers who mix into a compressor or equalizer as part of their approach are actually tailoring the mix to its sound and would not be served by removing it.

Multiple Mono versus Stereo Interleaved

Multiple mono refers to a stereo recording saved as two recordings, one of the left channel and one of the right channel, each in separate audio files. The term *stereo interleaved* means that both channels are combined into a single stereo audio file. Opinions vary about which is better. A few engineers are sure there is an audible difference and prefer multiple mono. By far the most popular belief is that there is no difference, and stereo interleaved mixes are preferred.

Selecting Stem Mastering

Stem mastering is mastering performed from submixes called *stems*. When all the stems are played simultaneously (e.g., drums, guitars, vocals, and bass), they make up the entire mix. This process is shown in Figure 3-1. When stems are provided, the mastering engineer has more control over the mix. Stem mastering is typically used for situations where productions have significant problems that are unable to

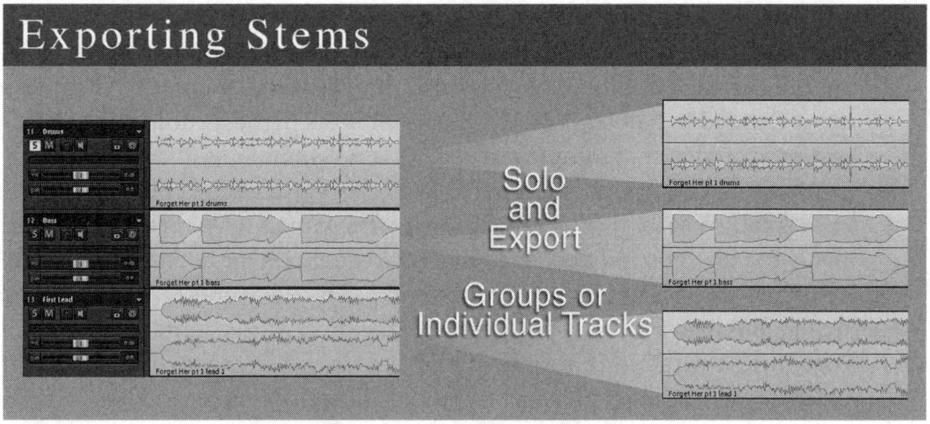

FIGURE 3-1   Solo an individual track or a group of tracks, and then export to create a stem. When all the stems of a mix are combined, they represent the entire mix.

be addressed in mixing or by conventional mastering. It also can be useful when a project needs cohesiveness that may not be possible due to various project restrictions. Stems may be required for some television/film/broadcast applications and are sometimes specified in recording contracts. Stem mastering requires extra time and a different way of working, and for those reasons, it is more costly. The vast majority of mastering engineers prefer to work from stems only when clients demand it. Very few mastering engineers prefer stems and have developed approaches for working with them.

Allowing Time for Mastering

Deadlines always should be clearly communicated to the mastering studio. Mastering can take time for developing tailored processing. Because of this, plenty of time should be allotted in the production schedule to permit the highest-quality mastering. Normally, two weeks is very safe, although under ideal conditions professional mastering sessions may take only a few hours.

Sample Rate

Mixes are often delivered for mastering at 44.1k; 48k; 88.2k; 96k; or 192k sample rates. In the digital audio world, samples are data points that make up the audio waveform and ultimately represent speaker movement. The term *sample rate* refers to the number of samples per second of a digital recording. There is some controversy surrounding the benefit of high sample rates. Critics quote the Nyquist-Shannon sampling theorem, which describes the relationship between sample rates and audio frequencies. To summarize the theorem: When the sample rate is divided in half (e.g., 44.1k), the result is the highest audio frequency that can be reproduced at that sample rate (e.g., 22.05 kHz).

Human hearing is thought not to extend beyond 22.05 kHz, so a sample rate of 44.1k is sufficient for the full range. There are scientists who have set up experiments that show that you may be able to sense frequencies far beyond this range. There are others who believe that the way the experiments were conducted render them invalid and who generally disagree with the proposition. Critics believe that when frequencies over 22 kHz are added to a recording, it leads to distortions within the audible range on most playback systems. These distortions are then interpreted as an improvement to the highest frequencies. They suggest that distortion can be added between 16 and 22 kHz to get the same or a similar effect. While the critics' argument about playback is not without merit, it may not take into account the digital-processing benefits within the audible range that can be experienced at higher sample rates during processing.

Digital processing can produce better results at higher sample rates. Today's digital processors often have internal upsampling features built in, at least somewhat negating the benefit of working with high-sample-rate recordings.

There are some advanced sample rate conversion algorithms that greatly reduce conversion problems to a possibly inaudible level.

Mastering to Tape

Some of the best mixes in the world are bounced to tape, and the mastering studio is provided with tape reels. Many legendary engineers still think that this is best and have their own favorite tape machines and calibration settings.

Bass and "Air"

The best mix is one with a perfect balance between the frequency ranges. Sometimes acoustics or monitoring do not permit an ideal balance, especially in home or project studios. When the ideal balance is not possible, most mastering engineers would agree that it is best to err on the side of too little bass and too little "air" (very high frequencies) in preparation for mastering.

Mix Problems

Mastering studios often have some method of providing amateur clients with information to improve their recordings outside the possibilities of mastering. This might be as simple as a list of resources for recommending to clients or as extensive as creating unique resources. Mastering engineers must maintain knowledge of audio concepts in order to provide assistance to mixing engineers. Most mastering engineers agree that technical mixing issues heard by the mastering engineer should be discussed with the client. Also, many mastering engineers think that it's best not to provide feedback except strictly on technical issues unless their opinion is sought.

Highly revered mastering engineers and studios are sometimes thought to achieve their sound only because they receive stellar mixes. This is untrue—the most talented mastering engineers usually can contribute something great even with less than ideal mixes and receive plenty of them.

CHAPTER 4

Accepting Mixes, Workflow, and Client Interfacing

You've prepared the mixes for mastering, and a mastering studio has been carefully chosen. Now it's time for the mastering studio to accept the mixes and get started. This chapter describes a few things to consider as the mastering begins. It also covers a few related topics that can be important throughout the mastering process.

Receiving/Importing Digital Recordings (Bit Rate, Sample Rate, and Levels)

When recordings are prepared for mastering, you must take special care to maximize their quality. Clipping should be prevented, and degrading conversions should be avoided.

Recordings are often requested from the client at 24-bit resolution at their original sample rate. Sometimes 32-bit Waveform Audio File Format (WAV) files are requested. In essence, 32-bit recordings are the same as 24-bit files except they include overload protection, preventing virtually all clipping problems. Requesting digital mixes at a peak level of –3 dBFS (decibels relative to full scale) or less allows the mixing engineer plenty of margin to ensure that clipping is not present. This is done without any perceptible sacrifice of quality when using the 24-bit format. Also, at 16 bits, a –3dBFS ceiling raises the noise floor more significantly than at the 24-bit rate, but it is not likely to cause any perceptible difference.

Data Integrity: CRC and Checksums

Although uncommon, errors can be introduced when copying and transferring digital files. To detect these errors, cyclic redundancy check (CRC) or checksum file verification can be used. Such verification is built into many operating systems, transfer algorithms, and file compression methods such as ZIP and RAR. There are many stand-alone CRC and checksum applications, such as the popular WinMD5 and Fastsum, which can provide assurance that file copies are exact. These programs generate a Message-Digest Algorithm 5 (MD5) code, and the code can be e-mailed. When the codes are received, they can be entered into a MD5 application that verifies the files to ensure that a perfect digital copy has been received. While this is an option, most professional mastering engineers do not perform this verification unless there appears to be a problem.

Requesting Information from the Client

As things are getting started, you need a few details about the client and the project. Information including artist name, address, contact and shipping information is requested. Also, you need the album title and other details to be placed in CD-Text. Often an order form is used to streamline the process. If MP3 files will be exported from the session, MP3 metadata will be requested, including cover art. CD-Text and MP3 metadata are discussed in greater detail in Chapter 12.

Receiving/Importing Analog Recordings

Often high-quality mixes will be delivered to the mastering studio on analog tape. This reel-to-reel tape size is typically ¼-, ½-, or 1-inch and rarely 2-inches. The mastering studio will need a calibrated and serviced tape machine for working with tape. Analog processing is usually performed as the analog source is recorded into the digital audio workstation (DAW), so only one digital conversion occurs. Sometimes the mastering studio is given both analog tape and a digital master, making it the studio's option about which to use.

Attended versus Unattended Sessions

Some people believe that mastering is best with unattended sessions because it is impossible for clients to be familiar enough with the mastering studio's monitoring system to make critical judgments. In an unfamiliar listening environment, clients could make a request that damages the ability of their mixes to translate to other systems. In contrast, attended sessions allow clients to understand more about the process, including the limitations. Attended sessions also can help to develop the relationship between the client and mastering engineer and make for easier communication.

Organizing Our Work

The organization of digital files is important for productivity. Naming conventions, folder/directory structures, desktop organization, backups, easy access to the original files, and file versioning systems are all very important. Mastering engineers should understand how their DAW works with digital files and take great care in keeping their work organized. Organization is especially important at the beginning and ending of a project.

Project notes and client details may be saved using dedicated software or inside the DAW. Magix Samplitude and Sequoia have an area where notes and client information can be saved along with the project. You also might use Google Docs, DropBox, or other online cloud services so that your information will be backed up automatically. DropBox is especially suited for this task because it allows for multiple past versions to be easily restored.

Saving/Copying/Pasting Processing Configurations

Organization doesn't stop with the files themselves. When a certain processing chain works well, it may be convenient to make note of the settings, or if plug-ins are being used, the configuration may be saved. In this way, you can recall the settings if later adjustments are necessary. Sometimes you may find something that doesn't work for the current project but may work well for a later project. Also, most DAWs have a feature where plug-in chains and settings can be saved into a *preset*. From there, adjustments can be made to suit each recording.

Importance of Timeliness

It is essential to give clients updates and information in a timely way. In the music industry, often deadlines and heavy schedules are involved. If time frames are not under control, problems can arise. Also, self-imposed deadlines, such as a listening party, can cause unnecessary stress. If circumstances affect the time frame of a project, it should be relayed to the client immediately. Communication is key to the relationship.

Each Recording Is Processed Individually

Each song has its own characteristics and identity. Mastering engineers process each song individually. It is usually best to completely reset your processors before moving on to the next song. You work so that the songs of an album or extended play (EP) sound their best and as though they belong together. This doesn't mean processing them the same way.

File Compression

When audio engineers hear the term *compression*, they think of dynamics processing. Compression also can mean digital *file* compression—a method of reducing the size of a digital file and, often, combining several files into one. The most common file-compression types are ZIP and RAR. To simplify file transfers, clients often compress the recordings of an album into a single file. This is also done to archive recordings for storage, minimizing necessary storage space. A few engineers believe that this process can have an effect on the quality of a recording, although it does not.

File-Transfer Options

Online file-transfer methods each have advantages and disadvantages. These services include FTP, WeTransfer, DropBox, YouSendIt, Proaudiobus, and Sendspace, among many others.

Focus on Customer Service

Educate, encourage, and empower. These are the main keys to customer service in mastering. Also, it can be the best decision to hold back opinions unless they are solicited by the client. Communication is usually best with an air of respect and positivity.

Time Frame for Revisions in Mastering

Making a revision (i.e., an edit or change) in mastering can take considerable time once the process is complete. This is so because entire tracks/albums must be processed, and the final output may need to be re-created. This also means that quality control must be performed again. Each of these takes time, which can quickly add up. This means that you should take precautions as you are getting started to avoid or reduce the possibilities of having to reprocess. You should be sure that you are working with the correct files and that you have complete information from your clients.

CHAPTER 5

Mastering Gear

S electing and understanding the equipment is a passion for most mastering engineers. It is a part of what makes studios unique. It's no wonder that the most popular online forum about professional audio is named "Gearslutz"! Mastering gear is usually chosen according to value, quality, character, and function. Some studios seek a balance between *color* processors and *clean* processors. Some want mostly clean or mostly color, whereas others focus only on gear that works well for a niche. This chapter gives perhaps the most complete coverage of the gear used in professional mastering available. The rare items and the common— you'll find them here.

Gear Lust

There is quite a bit of marketing influence, psychology, and herd mentality that affects gear choices. It is important to remember that while equipment is important, the greatest influence on the sound comes from monitoring, acoustics, technique, and skill.

Equipment Demos

Demoing equipment in the studio where it will be used is the best way to make purchase decisions. There are a variety of policies among dealers and manufacturers for arranging demos.

Stepped versus Continuously Variable Controls

Analog mastering gear will have one of two types of control knobs: stepped or continuously variable. *Continuously variable* knobs rotate smoothly across their range and have what feels like an infinite number of positions depending on how slightly the knob is adjusted. *Stepped* controls click into a fixed number of positions across their range and are normally preferred in mastering for repeatability.

Detented Potentiometers versus Rotary Switches

Continuously variable knobs are made of a component called a *potentiometer*. Some potentiometers are made with a *detent* or several detents (a type of step) that can be felt when turning the knob. These detents are not very reliable for precisely recalling settings. Instead, mastering-grade equipment often uses rotary switches with several stepped positions that can be selected. Rotary switches and detented potentiometers both have a number of positions across their range, but rotary switches allow for more precise values and a more solid electrical connection, which can result in a better sound. With rotary switches, you also can quickly and accurately recall settings. Potentiometers are notoriously inaccurate with a less solid electrical connection, which can negatively affect the sound. Also, this means that with potentiometers, you cannot set the left and right channels to the exact same values.

Analog Components

In a discussion about analog components, one might hear about vacuum tubes, inductors, transformers, op-amps, and switches. Usually these components are described in terms of their quality, character, transparency, or tactile feel.

Software Plug-ins versus Analog Processing

The comparison between analog and digital processing is completely different in today's world than in the past. Today it depends on the situation and task at hand. Most modern plug-ins are built with 64-bit architecture that allows higher quality than their 32-bit predecessors. This distinction is especially important for equalization in mastering. However, because of the complex nature of program material, plug-ins that may sound great for mixing are not suitable for mastering.

Opinions vary widely about the strengths of digital and analog processing. Analog equalizer designers believe digital equalization does not measure up to analog.

On the other hand, some digital equalizer designers believe digital can surpass analog. What are we to believe? The best thing is to trust your ears. You'll find that most experienced mastering engineers swear by their various analog equalizers, strongly believing they offer a more musical and diverse palette than digital equalizers. Among professional mastering engineers, analog is clearly preferred.

While analog equalizers are still generally preferred in mastering, the general consensus is that digital does EQ better than compression. However, it is almost universally believed that the quality of digital limiting actually has surpassed that of analog limiting. Also, it is worth mentioning that plug-ins are well suited for situations where there are extreme time constraints

Mastering Software Bundles

Mastering software is often sold in bundles. iZotope, Waves, Universal Audio, Plugin Alliance (Brainworx), Flux, and Sonnox have created popular mastering bundles.

Native Processing

In the audio world, *native* refers to an audio program or plug-in that does not require special accompanying hardware that helps to run the software. Digital Signal Processing (DSP) systems such as the UAD-2 have plug-ins that will only run using installed audio processing hardware that is therefore not native. If a plug-in requires a security device such as an iLok key, it is still considered to be native.

Retailers, Auctions, and Classified Ads

Mastering equipment is often purchased from Vintage King, eBay, Audiogon.com, Guitar Center Pro, and the for-sale section of the Gearslutz forum. Discover the approaches of Jaakko Viitalähde in "Connection and Calibration of an Analog Mastering Chain" in Chapter 16. Also see Robin Schmidt's "On Analog Multiband Compression and Audio Gear" in Chapter 16 for more about selecting mastering equipment.

Acoustic Treatment

Realtraps, GK Acoustics, and RPG Diffusor Systems are a few of the most noted manufacturers of acoustic devices for mastering studios. Acoustic devices one might use in a mastering studio include

- Clouds
- Vertical-well quadratic diffusers
- Array diffusers (also called Skyline under the RPG Brand)
- Cylindrical diffusers
- Pyramidal diffusers
- Absorption
- Bass traps

Monitoring Control Systems

Monitoring control systems allow for adjusting loudness, switching between sources, and other features. There is usually a zero position, so a calibrated reference loudness can be set. Often there is a useful DIM (dimmer) switch that allows for configurable loudness attenuation at the push of a button. Some feature input offset attenuation so that each input can be auditioned at matched levels. A few of the popular units include

- Crane Song Avocet
- Dangerous Music Monitor
- Dangerous Music Liaison
- Grace Design m905

Monitors (Speakers)

In recording and mastering studios, the speakers are known as *monitors*. They have a tremendous impact on the way you work and are considered to be the central tool of mastering. Without high-quality monitoring, your adjustments would not translate well to other playback systems.

Ideal Monitors

Monitors should be accurate, with a linear frequency response and extremely low distortion. Sound reproduction should be honest and unforgiving.

Passive versus Active Monitors

Active monitors have a built-in amplifier; *passive* monitors do not. Amplifiers have a significant impact on the sound, so different amplifier models should be auditioned with passive monitors to find the best match. The cost of a passive monitoring system is generally higher, with a higher maintenance cost and more challenging setup.

Full-Range versus Midfield versus Near-Field Monitors

Full-range monitors are the most common in mastering. Midfield monitors are also typical. Near-field monitors have a more narrow "sweet spot" and usually have less frequency linearity and accuracy. Almost without exception, near-field monitors are not considered to be suitable for mastering. Their nonlinear frequency response does not allow for accurate adjustments to be made. Also, the small sweet spot of near-field monitors is not convenient for attended sessions.

One of the advantages of full-range monitors is their wide sweet spot; also, their extended low-frequency response often makes subwoofers unnecessary. Some people say that mastering-grade full-range monitors are the biggest "secret" in mastering. When examining the equipment lists of legendary studios such as Masterdisk, Sterling Sound, and most others, high-quality full-range monitors are always there. Full-range monitoring opens a new world for most mastering engineers, although it is possible to satisfy some clients with near-field and midrange monitors.

For more on this topic, check out "The Case for Full-Range Monitoring" by Scott Hull in Chapter 16.

Common monitors used for mastering include

- Duntech Sovereign
- Dunlavy SCIV
- Tyler Acoustics Decade D1
- Nova Evolution and Evolution II
- Bowers and Wilkins 801s, 802s, and 803s
- Quested 3208 (Passive)
- Lipinski
- Barefoot Sound MM12s and MM27s
- Egglestonworks (including Andras and Savoys)
- ATC Loudspeakers

Upgrading to 5.1

5.1 is a speaker system with five speakers and one subwoofer and is used primarily in film/cinema. When upgrading to 5.1, full-range monitors are not often used for the entire system. For instance, three smaller Dunlavy SC-1s may be used with the large Dunlavy SCIVs. Bowers and Wilkins 805s might be used with the large 801s.

Crossovers

The most familiar crossovers are those built into many subwoofers. *Crossovers* are equalizer circuits that allow speakers of a multispeaker audio system to work within a specific frequency range. Crossovers are also inside most mastering-grade speakers. Usually, the subwoofer crossover frequency is adjustable. Typical subwoofer controls are shown in Figure 5-1.

General Placement

Monitors are generally placed in a free-standing way in mastering studios. This can be on stands or on the floor if they are large enough. Most engineers believe that speaker height is best at ear level, although some studios place them above the listening position, facing downward. Spikes, feet, foam, or other material may be used to decouple the speaker from the surface on which it rests. Otherwise, some of the speaker's energy could be transferred to the base. Placing speakers on a surface such as a desk or console is particularly problematic because reflections can be created that reduce accuracy.

Monitor Positioning

Some acousticians advise that each monitor should be set up with differences in distance from its closest walls. For example, the right monitor should have a different

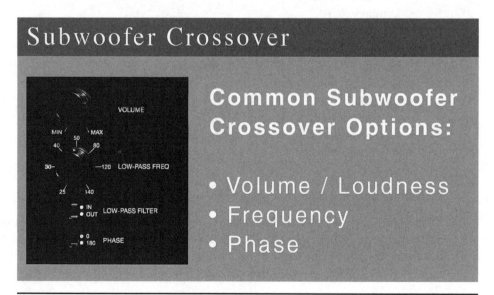

FIGURE 5-1   Mastering-grade subwoofers will have at least three controls: level, frequency cutoff, and phase.

distance between itself and each of the walls closest to it. The left speaker is set up in a way that mirrors the right speaker to have the same differences in distance from its closest walls. This is less of a factor in larger rooms because monitors are placed far away from walls.

Also, monitors are usually placed at an equal distance from both each other and the listener in a way that forms an equilateral triangle.

Subwoofers

When subwoofers are used in a mastering studio, there are almost always two, set up in a stereo configuration. Midfield and near-field monitors are virtually always used with subwoofers. Subwoofers are sometimes used with full-range monitors as well. Skilled acousticians can help with choosing the right subwoofers and with integrating them into the monitoring system. While it is best left to a professional, there are do-it-yourself calibration methods (discussed in Chapter 6). Subwoofer calibration depends on the crossover of the speaker, and mastering subwoofers require a configurable crossover point for integration with the monitors. Also, a crossover slope-adjustment feature can lend itself well to critical linear integration. The choice of subwoofer is not as critical as the choice of monitor. Many different kinds can be used successfully; therefore, my list of subwoofers is not exhaustive.

A few of the more common mastering subwoofers include

- Velodyne
- Bowers and Wilkins
- HSU Research
- M&K
- JL Audio

Monitor Amplifiers

A high-quality amplifier is essential for a mastering studios' monitoring system. The Pass Labs X-250 amp is becoming increasingly popular. Bryston amps have been standard for decades. There are many other great options considered to be mastering-grade. Mastering amplifiers are almost always class A, A/B, and lately D. Class D amplifiers have come a long way, with some being truly mastering-grade. They use less power and are often called *green* amplifiers. One general rule of thumb to gauge amplifiers is to compare their power ratings at 4 and 8 Ω. The closer the ratio is to 2:1 between the power ratings, the higher is the quality assumed to be.

Some mastering-grade amplifiers have a gain control built in; others don't. If there is no built-in gain control, an inline attenuator pad can be connected between the amplifier and the cable (this is usually going to be an XLR cable). High-quality inline attenuator pads are made by Crane Song, Schoeps, and a few others.

The most common mastering amplifier manufacturers are

- Pass Labs
- Bryston
- Cello/Mark Levinson/Red Rose Music
- Bel Canto
- Classe
- Krell
- Rotel
- Parason
- Hypex
- McIntosh

A/D and D/A Conversion

The term *analog* refers to some continuously variable quantity in the physical world. Examples of analog signals include electricity (voltage), sound, light, and speed.

Audio mastering ultimately deals with sound—an analog phenomenon. The speakers and amplifiers used to physically reproduce sound are analog.

Some conversion must take place to interact with analog devices using a digital system, such as a DAW. This conversion is accomplished with analog-to-digital (A/D) converters and digital-to-analog (D/A) converters.

Nearly all professional digital audio formats use a method called *pulse-code modulation* (PCM) to represent sound digitally. The most popular PCM format is the Waveform Audio File Format (WAV format). With PCM, audio is represented with samples, which are like tiny snapshots of a sound wave. Today's digital recordings involve thousands of samples per second; for example, at the CD standard 44.1k sample rate, there are 44.1k samples per second. Each sample is placed at some level of intensity, which is called a *bit*. Figure 5-2 shows how bits and samples are usually represented.

Analog-to-digital (A/D) conversion involves sampling an analog signal at a time-constant interval and quantizing a sample value to the closest value of the digital range. Digital-to-analog (D/A) conversion involves reconstructing an analog signal from the digital signal.

There are three main converter functions in a mastering studio—monitoring, reproducing, and recording. To feed the monitoring system, a *monitoring* D/A converter is used. Monitoring D/A converters can be stand-alone or built into some monitoring control systems, such as the Crane Song Avocet. A monitoring

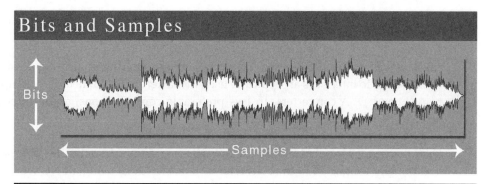

FIGURE 5-2   How bits and samples are represented in the visualization of a waveform.

D/A converter is separate from a *reproducing* D/A converter (sometimes called a *pitching* D/A converter), which converts the signal coming from the DAW to analog for processing. After analog processing, the signal goes to the *recording* A/D converter (sometimes called a *catching* A/D converter), which sends the analog processed audio to the DAW to be recorded. If the mastering source is analog tape, the pitching D/A converter is not necessary, which minimizes the number of conversions.

Characteristics of a good mastering converter include low signal-to-noise ratio, low distortion, and low/suppressed jitter. The quality of these characteristics is the result of analog circuitry design, converter-chip quality, and clock design. The converter chips in A/D and D/A converters are sometimes mistakenly thought of as playing the greatest role, but their impact is actually small in the overall quality of the conversion. It is the analog circuit architecture of a converter system that plays the most significant role. For example, many sigma-delta circuit types are becoming preferred over ladder network types owing to their minimal distortion.

High-quality mastering converters are manufactured by:

- Lavry
- Crane Song
- Prism
- Benchmark
- Apogee
- Antelope Audio
- Burl
- DCS
- Forssell
- Lynx
- Weiss Engineering

Audio Interfaces

Audio interfaces are hardware that connects a computer to external audio devices or sources. Audio interfaces are typically used in mastering for connecting a computer to A/D and D/A converters. In less preferable configurations, audio interfaces serve as both interface and converter.

In mastering studios, the audio interface is typically connected to an A/D or D/A converter by an AES/EBU or S/PDIF connection. Usually the interface is additionally connected to the converter or a master clock using a BNC wordclock connection and is slaved to the converter as an external clock source. The basic mastering studio connections involving the A/D or D/A converter are shown in Figure 5-3.

To perform analog mastering from a digital source, an audio interface must have at least one digital input and output (S/PDIF or AES/EBU). Ideally, BNC wordclock connections are used for syncing the interface to an external clock. It is possible to sync a wordclock signal over S/PDIF or AES/EBU connections instead of BNC. However, depending on the manufacturer's design, syncing over AES/EBU potentially could be more prone to jitter problems.

PCI or PCIe format interfaces, such as the one shown in Figure 5-3, are often preferred over USB or FireWire because of generally higher stability, although some USB and FireWire devices are truly just as stable.

For mastering interfaces, perhaps the most significant attributes are the quality of an interface's PLL lock, driver stability, and the number of digital inputs and outputs.

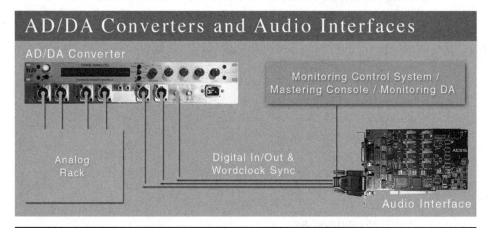

FIGURE 5-3 How an A/D or D/A converter, analog rack, monitoring control system, and audio interface are interconnected.

High-quality audio interfaces used for mastering are manufactured by:

- RME
- Lynx
- Weiss Engineering

Wordclocks/Master Clocks/Distribution Amps

A digital audio signal is made up of thousands of individual samples. Wordclocks are used for timing the recording or playback of these samples. Wordclocks are judged on how evenly they accomplish this task.

Inaccurate clocks produce jitter. *Jitter* can be loosely defined as anomalies as a result of inaccurate digital signal timing.

Master clocks and distribution amps are all devices associated with wordclocks. Master clocks are dedicated wordclock units that can supply a wordclock signal directly to many digital devices.

Wordclocks are almost always built into A/D and D/A converters and audio interfaces. Most often, mastering studios use the built-in wordclock of their mastering-grade A/D and D/A converters. All digital devices of the studio are synced to this single clock.

Wordclock signals can be synced between devices using BNC, AES/EBU, or S/PDIF connectors. This is called *wordclock distribution*. BNC connections are the ideal wordclock connection, producing the least amount of jitter. BNC connections usually operate at 75 Ω and require 75 Ω cables with BNC connectors.

When a master clock is used, each unit is connected directly from each WC OUT of the master clock to the WC IN of the digital units. When an A/D or D/A converter's wordclock is the master wordclock, there are a few options for making the connections. The most popular option involves using a T-connector to daisy-chain each unit in the chain. The T-connector is never to be used as a splitter. A T-connector is placed into the WC IN of each digital unit, and each unit is connected to the next (daisy-chained). In this configuration, if the last digital device in the chain is not *self-terminating*, then a 75-Ω terminator is connected into the open end of its T-connector.

The next option is to use a distribution amp. Wordclock distribution amps split a master wordclock signal into multiple wordclock signals for syncing multiple devices. If there are only two digital devices (e.g., an A/D converter and audio interface), then the WC OUT of the master wordclock can be connected directly to the WC IN of the other digital device.

There is some controversy regarding which BNC wordclock distribution methods provide the best result; although each of these methods is acceptable and can be found in use by major mastering studios.

Wordclock signals also can be synced over AES/EBU and S/PDIF connections. This is normally seen as inferior owing the possibility of to added jitter. Because of this, mastering studios typically opt for BNC connections to carry wordclock signals.

With some digital devices, the synchronization must be set up using software or the device itself. User manuals normally provide information for wordclock syncing and indicate whether a device is self-terminating.

A couple of examples of wordclocks, master clocks, and distribution amps include

- Atomic master clock
- Antelope Isochrone 10M and Trinity
- Big Ben

Analog Equalizers

In mastering, equalizers are used for many tasks, including adjustment of tonality and correction of frequency problems. Analog equalizers are especially suited for adjusting tone. They are seen as having a more diverse sound between different models than digital equalizers and as imparting a more pleasing quality.

Mastering equalization involves several equalizer filter types including: shelving, parametric/bell-shaped, high-pass and low-pass. Mastering studios seek out preferred filters for each type.

Equalizers have strong and weak points. For example, some are perfect for making narrow cuts, others are great for adding "air," whereas others excel at wide tone shaping.

Selecting an analog equalizer is about finding the combination that works for you. Forums are great places to find common perceptions about various equalizers, but trying them out is essential for evaluation. If a mastering studio will have only one unit, a versatile unit should be selected. If a mastering studio will have several units, it is important to make selections with complementary strengths.

Commonly used mastering equalizers include

- Buzz Audio REQ-2.2
- Sontec (various models)
- Crane Song Ibis
- Milennia NSEQ-2 and NSEQ-4
- Prism Sound Maselec MEA-2 Mastering Equalizer
- Dangerous Music Bax EQ
- Manley Massive Passive
- Great River MAQ-2NV

- Gyraf Gyratec XIV
- Focusrite Isomorphic Equalizer Blue 315
- Chandler TG12345 Curvebender
- Bettermaker 232P
- EAR 825/822Qs
- GML 8200 and 9500
- SPL Passeq
- API 5500
- Weiss Engineering equalizers
- Pulse Technologies Pultec EQP-1A3
- Manley Stereo Pultec EQP1A
- Manley Mini Massive
- A-Designs EM-EQ2
- TK Audio Tk-lizer
- Inward Connections DEQ-1
- Mercury EQ-P1
- D W Fearn VT-5
- Knif Soma
- Summit Audio EQ-200
- TFPRO P9
- Kush Audio Clariphonic Parallel Equalizer
- Neumann mastering equalizers

A few custom equalizer designs are popular among mastering engineers. Custom equalizers built for mastering studios include

- Davelizer
- Barry Porter BPEQ

Plug-in Equalizers

Like most processors, digital plug-in equalizers are getting better all the time. Some of their features are comparable with those of analog equalizers, although, in mastering, digital plug-in equalizers are not yet viewed by professionals as on par with their analog counterparts.

The most commonly discussed weakness of plug-in equalizers is their ability to make high-frequency adjustments, including adding "air." Analog equalizers are almost universally viewed as superior for these tasks.

Minimum-phase is the most popular plug-in equalizer type in mastering. Some operate only in minimum-phase mode, whereas others have a minimum-phase option among other modes. Minimum-phase is viewed as being the most musical-sounding digital equalizer option.

Popular digital plug-in equalizers for mastering include

- DMG Equilibrium
- Fabfilter Pro-Q
- Sonoris
- Flux Epure
- Algorithmix Red
- Refined Audiometrics PALPAR EQ
- Sonalksis
- Mellowmuse EQ3V

Dynamic Equalizers

Dynamic equalizers activate their filters depending on the content of the mix. Perhaps the most popular dynamic equalizer for mastering is the Weiss Engineering EQ1-DYN.

- Weiss Engineering EQ1-MK2 and EQ1-DYN
- Brainworx bx_dynEQ
- Sonnox Oxford SuperEsser

Compression/Expansion

Compression and expansion are processes that act on the loudness of a sound. Compression lowers the loudness according to its settings, and expansion raises the loudness. Let's find out more about these processors.

Compression

The most well-known dynamic-range processing used in mastering is compression. With compression, peaks are reduced according to the compressor's settings.

Expansion

Expansion is used in mastering to increase the dynamic range and is built into some compressors. With this processing, the peaks are expanded according to settings. Most mastering engineers rarely use it.

Parallel Compression

With parallel compression, the softest sounds are raised, and the peaks are not affected. This type of compression is most popular in acoustic and orchestral mastering. A compressed signal (usually heavily compressed) is mixed in with the original. This is accomplished effortlessly by using a compressor with a "Mix" knob feature. It also can be accomplished manually with a DAW or with a mastering console that includes a "Mix" feature.

Side-Chain

Many compressors have a *side-chain* feature that allows the compressor's action to be based on a signal other than the one being processed. In mastering, sidechains are sometimes fed a copy of the same signal being processed but with the bass frequencies de-emphasized. In this way, the compressor's action responds less to bass frequencies.

Selecting a Compressor

Compressors are often selected based on their character. Compressors might be described as *transparent* and *clean* when they have little or no effect on the sound. Other compressors impart a distinct sound of their own, called *character* or *color*. Mastering studios usually have options for both transparent and color compression.

Analog Compressors

Analog mastering compressors are built with various gain-control elements. The types include VCA, opto/ELOP, variable mu, FET, and PWM. The term *voltage-controlled amplifier* (VCA) describes virtually all analog compressors, although the term has become accepted as referring only to those using integrated circuits (ICs). VCA compressors have a widely varying sound and are often very fast. Opto compression is often thought of as providing a clean sound, imparting a "glue" effect, and is typically slower. Variable mu compression is often said to add a thick, warm, and rich character. FET compression is not widely used in mastering. Pulse-width modulation (PWM) is typically used to emulate other types of compression.

Analog mastering compressors have different actions, such as root-mean-square (RMS) sensing, peak sensing, and similar variants. RMS sensing compressors' action is based on the average level; peak sensing action is based on peaks.

The hardware details are good to know and may help in making a selection. While this is true, the details mean little about how a compressor might sound. The sound of a unit can only be evaluated by listening.

These are the most common mastering compressors:

- API 2500 (VCA)
- SSL 4000 Series (VCA)
- Neve 33609 (VCA)
- dbx 160SL Blue Series (VCA)
- Elysia MPressor (VCA)
- Elysia Alpha (VCA)
- Vertigo VSC-2 (VCA)
- Foote Control Systems P3S—Typical modification: External sidechain equalizer connected to it (VCA)
- Foote Control Systems P4 (VCA)
- SSL X-Logic G-Series (VCA)
- GML 8900 and 2030 (VCA)
- Roll Music Systems 755 (VCA)
- Dramastic Audio Obsidian Compressor (VCA)
- NTP 179-120 (VCA)
- Rupert Neve Designs Portico II Master Buss Processor (VCA)
- SPL Kultube (VCA)
- Shadow Hills Mastering Compressor (opto/VCA)
- Maselec MLA-2 (opto)
- Pendulum OCL-2—Typical modification: Stock tubes replaced with New Old Stock (NOS) GE 6072A tubes and external sidechain equalizer connected to it (opto)
- Manley Vari-Mu—Typical modification: Sidechain or M/S features are added by Manley. The most sought-after version is the older version with 6386 tubes (variable mu)
- Drawmer S2 (variable mu)
- Drawmer S3 (opto/variable mu)
- Thermionic Culture Phoenix (variable mu)
- Fairchild 670 (variable mu)
- Pendulum ES-8 (variable mu)
- Esoteric Audio Research 660 (variable mu)
- Tube Corporation SR71 Blackbird (variable mu)
- QES Labs Variable GM (variable mu)
- Cartec THC (Tone Harmonic Compressor) (variable mu)
- Gyraf Gyratec X (variable mu)
- Drawmer 1968 (FET)
- Daking Fet II/Fet III (FET)
- Crane Song STC-8/M (PWM)
- Crane Song Trakker (PWM)
- Dave Hills Designs Titan (PWM/VCA)
- DW Fearn VT-7 (PWM)

Plug-in Compressors

Plug-in compressors have come a long way in terms of the quality they provide. Today some are suitable for mastering tasks, although analog compressors are still preferred by professional mastering engineers. The number of plug-in compressors that claim to be mastering-grade are too numerous to list here. Listed below are some of the most popular and highest quality.

- Magix Am-munition
- PSP Mastercomp
- Fabfilter Pro-C
- DMG Audio Compassion
- Flux Solera II/Alchemist
- Brainworx Vertigo
- Elysia MPressor
- Elysia Alpha
- Waves API 2500
- Waves and UAD SSL 4k
- UAD Precision Comp
- Mellowmuse CP3V
- Steven Slate FG-X Compressor Section

Multiband Compressors

Multiband compression allows compression to be applied only to specific frequency bands.

Analog Multiband Compressors

Analog multiband compressors include

- Prism Maselec MLA-3
- Tube-Tech SMC-2BM

Digital Multiband Compressors

There are a vast number of digital multiband compressors. A few of the most popular include

- TC System 6000 MD4
- Waves C4
- Waves LinMB
- T-Racks
- iZotope Ozone 5
- UAD Precision Multiband

Maximizers/Limiting

Digital limiters, sometimes called *maximizers*, are an important tool when the client needs a loud master. Over the years, digital limiters have nearly replaced analog limiters because of their precision and quality.

There are very many on the market, so only the top tier is listed:

- Fabfilter Pro-L
- Izotope Ozone 5 Limiter
- Voxengo Elephant
- Slate Digital FG-X
- Waves L2

Stereo/Mid-Side Processors

Stereo processing for mastering usually involves some form of mid-side processing or delay effects.

What Is Mid-Side?

Mid-side (M/S) refers to a two-channel recording made up of a middle and a side component instead of left and right. The *middle signal* is the audio present in both the left and right speakers of a stereo recording. The *side signal* is the audio that is in either the left or the right speaker but not in both. Essentially, middle is all the mono audio, and side is stereo audio.

Mid-side is also called *sum-difference* because the middle channel is the sum of both the left and right channels (L + R), and the side channel is the difference (L − R). The various synonymous terms used for middle and side are shown in Table 5-1.

Most stereo-to-mono conversions in the consumer and professional audio world are actually creating the middle channel because it is simply a sum of the left and right channels.

In the days of vinyl, mid-side was called *lateral-vertical*. The mono information was the side-to-side (lateral) movement of the needle groove, and the stereo/side information was the up-and-down (vertical) movement of the needle groove. Also, the useful technique of converting a stereo recording to mid-side, performing equalization processing, and then converting back to stereo is called *shuffling* or *Blumlein shuffling*, named after Alan Blumlein.

TABLE 5-1   Synonymous Terms for Mid-Side

Mid	Mono audio/M/sum/L + R/lat/lateral
Side	Stereo audio/S/difference/L − R/vert/vertical

Stereo Processors

Stereo processors that one may find in use in professional mastering studios include

- Rupert Neve Designs Portico II Master Buss Processor
- Bedini BASE
- M/S features are built into several mastering consoles

Analog M/S Converters

Mid-side processing is growing increasingly popular in the analog domain. It requires the use of a processor with mid-side built in or a mid-side converter such as these:

- Avenson Audio Mid-Side
- Dangerous Music S&M
- SPL M/S Master
- Vertigo Sound VSM-2
- D.A.V. SIPP
- Alice Sum & Diff Pak
- DIY KA Electronics MS Matrix

Digital Stereo Processing

Digital stereo processors are popular digital tools in mastering for stereo and mid-side processing:

- Algorithmix K-Stereo by Bob Katz
- DDMF Metaplugin
- Sonalksis Stereo Image Processor
- IZotope Ozone Multiband Stereo Processor
- Matthew Lane's DrMS
- Brainworx XL
- Brainworx bx_digital

- SHEPPi
- Zynaptiq Unveil

De-Essers

De-essing reduces the harshness of sibilant sounds. Sibilance is the "*s*" sound in vocals. De-essing is a process that is very well suited for using in M/S mode, on the middle channel, because that is where vocals typically sit. While some mastering studios use outboard de-essers, others use plug-in de-essing or no de-essing at all.

Analog De-essers

Popular analog de-essers used in mastering include

- Maselec MDS-2
- Empirical Labs DerrEsser

Digital De-essers

A few of the most well-respected digital de-essers used in mastering include

- Weiss Engineering DS1-MKII
- Digitalfishphones Spitfish
- Sonnox Oxford SuprEsser
- Sonalksis DQ-1 & CQ-1
- SPL De-esser

Restoration and Noise Reduction

When recordings have noise, hiss, clicks, or pops, then restoration or noise-reduction processing may be needed. Popular mastering restoration/noise-reduction packages include

- iZotope RX
- Cedar Retouch
- Sony SpectraLayers Pro
- Weiss Engineering DNA1
- Algorithmix ReNOVAtor
- Waves Restoration Bundle
- Sonnox Restore Bundle

- Zynaptiq Unfilter
- Magix Spectral Cleaning and Restoration Suite

Harmonic Enhancement and Saturation

Professional mastering engineers often decry harmonic enhancement processes and for good reason. These processes can easily do more harm than good, so they must be used carefully. Digital processing is improving every day, and there are now options for harmonic enhancement and saturation that sound good. The Sonnox Oxford Inflator has become one of the most popular digital tools for this task.

Analog

These two analog harmonic enhancement and saturation devices are among the most popular:

- Anamod ATS-1
- Thermionic Culture Vulture

Digital

Digital harmonic enhancement and saturation have come a long way in the past few years. A few of the most popular plug-ins include

- Sonnox Oxford Inflator
- Slate Digital VCC
- Crane Song HEDD-192
- UAD Ampex
- UAD Precision Maximizer
- Mellowmuse CS1V
- Mellowmuse SATV
- Waves Kramer Master Tape
- Sonimus Satson
- Plugin Alliance Noveltech Character
- IZotope Ozone Multiband Harmonic Exciter

Routers/Patchbays

Routers and patchbays are used to configure the number and sequence of devices being used. They often look like rows of inputs and outputs and allow convenient switching. *Patchbays* (originally called *jackfields*) are the simplest and most cost-effective choice for this task. Devices are connected to the patchbay, and patch cables are used to interconnect devices.

Digital *routers* allow settings to be saved and provide fast configuration. All devices are connected to the digital router, and then configurations are made with software or menu controls. Some digital routers are *asynchronous*, meaning that they can work with a variety of signals and sample rates. Synchronous routers do not allow a variety of sample rates because they are synced to a digital clock to which all devices must be synced.

Patchbays come with various connections and usually have at least *normal*, *half-normal*, and *full-normal* options. XLR patchbays set without normals are the standard for use in mastering.

Although some engineers have reservations about using patchbays, they can be found in use by many of the most respected studios, including Sterling Sound.

A wide variety of patchbays is available, so this list is limited to the top tier:

- Z-Systems
- Metric Halo
- Mosses & Mitchell
- SSL X-Patch
- Smartpatch ARC32 Patchbay

Consoles

Mastering consoles are similar to routers, although they allow many more options and are specifically geared toward mastering workflows. There are several benefits of working with a mastering console. One of the biggest benefits is the ability to instantly switch between devices for comparison. With some, it is possible to effortlessly change the sequence of devices. Another benefit is control over the input and output levels so that they may be optimized for each analog device. Some analog devices have a sweet spot, and some consoles allow working at those optimal levels. Most make it possible to audition the original recording and the analog processed version at a matched loudness level for making a comparison.

Selecting a mastering console is all about the options. The choice depends on the desired workflow and the devices in use.

- SPL MMC1
- SPL MasterBay S
- Crookwood M Series
- Maselec MTC-2
- Manley Backbone
- Dangerous Music Liason
- Dangerous Music Master

- Shadow Hills Industries Equinox
- I.J. Research K-1
- Little Labs Digital Audio Mastering Router
- JLM Audio JLM Mastering Console
- Muth Audio Designs consoles (classic and not currently in production)

Headphones

Headphones are used in mastering primarily for quality control. Headphones can help to listen without the effects of room acoustics and can be very useful when the closest detail must be heard. Virtually every mastering engineer agrees that processing adjustments or decisions should never be made using headphones.

Common mastering headphones include

- Sennheiser HD800, HD600, and HD580s
- Grado
- Beyerdynamic

Acoustic Environment Simulation for Headphones

There are several inherent problems with headphone monitoring. Bass frequencies cannot be reproduced properly in headphones. Also, the perception of the stereo field is much different from listening to speakers in an acoustic environment. Reflections occur in an acoustic environment, and each speaker is heard with both ears. There are a few digital plug-ins that help headphone monitoring overcome these issues to some degree. Even with these plug-ins, mastering with headphones is not practiced by any professional mastering engineer.

- ToneBoosters TB Isone
- Mildon HC38
- 112dB Redline Monitor
- Acudora HDph2X
- Refined Audiometrics HDPHX

Digital Audio Workstation Software

A computer and software system used for recording audio is called a *digital audio workstation* (DAW). Often just the software itself will be called a DAW. Sometimes software comes along with specific hardware, as with Pyramix and SADiE (although SADiE now offers a native version).

Generally accepted mastering DAWs are listed below in approximate order of popularity:

- Magix Sequoia
- Magix Samplitude
- SADiE
- Wavelab
- Pyramix
- Sonic Studio Soundblade
- Audiocube
- DSP Quattro
- Waveburner
- Peak

Playback Sources

For mastering studios, it is good to have a variety of playback systems for importing sources. Such systems should be carefully selected and maintained.

- Analog tape machine (The Studer A80, A810, and A820, as well as the Ampex ATR-102, are among the most popular used for mastering.)
- DSD
- DAT
- ADAT
- Cassette

DDP Software

The Disc Description Protocol (DDP) is a computer data format that contains all the information for creating a CD. Duplicators/replicators can use DDP files to manufacture CDs instead of copying an actual CD. Also, mastering clients can audition a mastering studio's work by listening to a DDP with specialized software. This allows clients to preview all aspects of their CDs, including pause times and CD-Text info. Some DAW software will export DDP files; some will not. The latest DDP software offers new options that are interesting to almost any mastering engineer.

- Tone Proper Software's Tonic—Client DDP preview, time-stamped notes similar to Soundcloud
- Sonoris DDP Creator—Client DDP preview
- Cube-Tec DDP-Solution

- Hofa CD-Burn & DDP
- Audiofile Engineering Backline—DDP Player for Mac

Sample-Rate Converters

Often sample-rate conversion in mastering is from a high sample rate (such as 88.2k, 96k, or 192k) to the CD standard sample rate of 44.1k. Sometimes the conversion is to the DVD sample rate of 96k. Sample-rate conversion quality has an audible impact on the sound of a recording. Because of this, high-quality sample-rate conversion is used in mastering. There currently is a great comparison of sample-rate converters at http://src.infinitewave.ca/.

Software

Software sample-rate conversion has developed tremendously, with some meeting or exceeding the quality of highly revered hardware converters. The software below performs the highest-quality conversions according to the Infinitewave comparison:

- FinalCD 0.12
- IZotope RX Advanced V2 and 64-bit SRC
- Voxengo r8brain Pro Linear Phase

According to this comparison, FinalCD 0.12 has the very lowest distortion of all sample-rate converters, hardware and software alike.

Hardware

These hardware sample-rate converters are seen as mastering-grade:

- Weiss Engineering SFC-2
- Z-Systems z-2src, z-3src, and z-8src
- RME ADI-192 DD and ADI-192
- Lucid SRC9624
- Mytec Stereo192 SRC

5.1 Mastering

Mastering for 5.1 audio involves a much higher investment than stereo mastering. Monitoring, amplifiers, acoustics, and processing must be considered. For cost and practical reasons, 5.1 mastering is often done with digital processing. Such digital processors as well as monitoring control systems are listed here:

- TC System 6000 (complete solution for 5.1 processing)
- ZSYS z-Q6 and z-CL6 (complete solution for 5.1 processing)
- Crane Song Avocet 5.1 Expansion (5.1 monitoring control system)
- SPL Monitor Control System
- Crookwood Monitor Control System

Meters

While meters have some use in mastering, they indicate almost nothing about how to make adjustments to the tone of a recording. In mastering, one should use ears, not eyes. With this in mind, meters can help to provide a quick visual reference and some useful information.

Spectrogram, peak, RMS, LUFS, and bit meters are typical in mastering studios. Spectrograms use color to show the intensity of frequency energy across the spectrum. Peak meters show the level of waveform peaks. RMS meters are slower than peak meters and show the average levels, which are closer to how the human ear perceives loudness. LUFS meters are the closest meter we have to a representation of the human perception of loudness. Bit meters show the bit rate of the recording (e.g., 24 or 16 bits). Popular mastering meters and visualizations include the following.

Analog

Some mastering engineers delight in their analog meters, including:

- Dorrough
- VU (Weston's are very popular, among others)

Digital

Popular digital meters used for mastering include

- Logitek Ultra-VU (digital hardware)
- Waves PAZ
- DK Audio
- SpectraFoo
- Voxengo SPAN

- K-Meter (K-12, K-14, and K-20)
- Stock meters included with DAW

Mastering DSPs

Mastering digital signaling processors (DSPs) are hardware units that perform digital audio processing. All have corresponding software except System 6000, which has a built-in display.

- UAD-2
- TC System 6000
- TC Finalizer
- TC Powercore (discontinued but still supported)
- Duende (not commonly used for mastering)
- Waves APA (discontinued and unsupported)

Combination Processors

There are some hardware units that have multiple mastering effects built into one unit. A few of these include

- Rupert Neve Designs Portico II Master Buss Processor (stereo processing/compression)
- Legendary Audio Masterpiece by Rupert Neve
- TL Audio Ivory 5052 MK2 (equalizer/dynamics/compression)
- SPL Vitalizer (harmonics/bass enhancement/stereo processing/equalizer)

Metadata-Embedding Software

Mastering engineers are often required to embed metadata into several file types. This list features a few popular programs for embedding metadata in lossy and nonlossy file formats.

- Jaikoz Audio Tagger
- Mp3tag
- Mediamonkey
- Quesosoft BWAV Writer

Forensic Audio Software

Performing audio processing for forensic analysis normally involves restoration and noise reduction. However, there are applications for *voice biometrics* that help the user to determine if a person's voice would be capable of making the sounds heard in a recording of a voice. Law enforcement uses voice biometrics to exclude suspects. Easy Voice Biometrics and Nuance LVIS-Preforensic are two popular software applications for this task. Each year, the Biometric Consortium Conference and Technology Expo is presented in the United States to showcase biometric technology.

Connections and Cables

There are several different kinds of connections between devices that one may find in a mastering studio. Connections have two components, the connectors and the cables. The effect of cables is often exaggerated, although cables are certainly part of the chain and can have an impact. Generally, shorter cable lengths provide higher quality and fewer problems. Also, the number of connections is best kept to a minimum. The most popular brands are Belden (including the very popular 1800F AES), Mogami, and Canare.

Analog XLR

Analog XLR cables are common for analog connections in mastering studios. The XLR cables used in studios are balanced connections with three pins, carrying a ground, a positive signal, and a negative signal. In mastering studios, the connections between devices are line-level connections. There are two common line-level connection standards. One called +4 is used for professional audio, and −10 is typically used for consumer audio. There is also the less common +6, which is a German standard. These line-level standards designate the average and peak voltage for the connection. The various line-level standards are shown in Table 5-2.

TABLE 5-2   Line-Level Standards

Usage	Line Level		
	Nominal Level	**Nominal Level (VRMS)**	**Peak Amplitude (VPK)**
Pro Audio	+4 dBu	~1.228	~1.737
Consumer Audio	−10 dBV	0.316	0.447
Germany	+6 dBu	~1.550	~2.192

AES/EBU

AES/EBU is the abbreviation for Audio Engineering Society/European Broadcasting Union; it is the most common digital connection in mastering studios. AES/EBU cable is 110 Ω, 24 bits, balanced, and uses XLR connectors. Wordclocks can be synced over this connection. Virtually every audio engineer considers it to be the highest-quality digital audio connection. The maximum recommended length is typically 300 feet.

S/PDIF

S/PDIF is the abbreviation for Sony/Phillips Digital Interface (IEC 958); it is a digital connection. S/PDIF cable is 75 Ω, 16/24 bits, and unbalanced. Wordclocks can be synced over this connection. The maximum recommended length is typically 32 feet.

1/4th, 1/8th, and Minijacks

1/4th, 1/8th, and minijacks can be balanced using a connector called *tip-ring-sleeve* (TRS) or unbalanced with a connector called *tip-sleeve* (TS). TRS connectors are often called *stereo cables*, and TS connections are called *mono*. They are familiar to almost all audio engineers.

Multi-Pin Connectors

One may encounter multi-pin connectors in some mastering studios. These are often used with mastering consoles and for stereo linking (as with the Dave Hill Designs Titan). The most common in mastering studios are Elco/EDAC (20, 38, 56, 90, or 120 pins) and D-Sub (9, 15, or 25 pins). One may rarely encounter DT12/ Cannon FK-37 (it is the highest-durability multi-pin connector, 37 pins), DB25 (25 pins), and MASS W4s (176 pins), all of which are more common for live and television audio. Modifying the connections with these connectors is performed using insertion, extraction, and crimping tools.

RCA

Analog RCA connections are most common with consumer equipment and are unbalanced connections. In professional audio, RCA connectors are used to carry digital S/PDIF signals.

BNC Wordclock

BNC Wordclock connections are 75 Ω, with BNC connectors and typically RG-58 (lower quality) or RG-213 (higher quality) cable. The maximum recommended length is typically 50 feet.

Bantam/TT

Bantam/Tiny Telephone (TT) connections are 4.4-mm (0.173-inch) plugs, and they may be balanced (TRS) or unbalanced. Most often these connections are used with patchbays, although typically mastering studios use XLR patchbays.

ADAT Lightpipe

ADAT Lightpipe is fiber-optic cable with optical connectors. It can carry up to eight channels of 24-bit audio at a 48-kHz sample rate. Wordclocks can be synced over this connection. The maximum recommended length is typically 30 feet.

USB

USB connections are very common for connecting devices to a computer. The earliest USB 1.0 connections supported a data-throughput rate of 12 Mbps (megabits per second). When USB 2.0 was released in 2000, the throughput was increased to 480 Mbps. Finally, in 2008, USB 3.0 was released, increasing the speed to 5 Gbps (gigabits per second).

FireWire

FireWire (also called *IEEE 1394*) is another connection used for connecting devices to a computer. Some audio interfaces connect using FireWire. The original release of the FireWire standard was called FireWire 400. It supported speeds up to 392 Mbps, with a maximum cable length of 14 feet. The next release was FireWire 800, which supported much faster speeds of 3,200 Mbps.

Thunderbolt

Thunderbolt (also called *Light Peak*) is a digital connection that uses either a copper (10 Gbit/s) or optical (20 Gbit/s) cable and a Mini DisplayPort (MDP) connector. These connections are similar to PCIe connections without having to install hardware inside a computer's case.

PCI/PCIe

Peripheral Component Interface (PCI) is a standard for connecting hardware devices inside a computer. The original PCI slots run at a maximum speed of 533 MB/s (megabytes per second) at 64 bits and 266 MB/s at 32 bits. PCI was followed by PCIe, which has several speeds, including 1x, 2x, 4x, 8x, and 16x. Faster slots can support lower-speed cards. There is also PCI 1.0/1.1, 2.0, and 3.0. Each of the speeds and versions has different rates, all of which are very fast.

CHAPTER 6

Mastering Acoustics and Monitoring

Acoustic science is a vast enough subject to be covered by several books. Because of this, the focus here will be on the most important acoustic-related topics associated with mastering studios. There are a few generally accepted concepts, although each engineer's room has its own considerations.

Hiring a Professional Acoustician

When the time comes to open a new mastering studio, it is generally recommended to hire a trained acoustician, acoustic engineer, or architecture firm with experience in acoustic design. A mastering studio may begin from a new structure, or an existing building will be gutted and a new interior will be designed and constructed. Businesses that offer services in this area can be found with a search for architectural acoustics, acoustic consulting, and acoustic engineering. The specialists who work with acoustics are called *acoustic engineers* or *acousticians*. They can help to design rooms with a balanced frequency response or improve an existing room with acoustic treatments. Some of the common acoustic treatments found in mastering studios are described in this chapter.

Room Modes

Room modes are the collection of resonances that exist in a room when acoustic energy is present. Room modes are the main reason for acoustic treatment.

Absorption

Various materials are used to absorb problematic sound-wave reflections. One of the most popular absorption materials is acoustic fiberglass, with Owens and Corning 703 being the standard. Owens and Corning 703 is rigid, making it easy to work with. Mineral wool and ultratouch cotton are alternatives, although neither is rigid. Too much absorption in a room can lead to a dead sound that is difficult to work with for extended periods. Mastering engineers often prefer a listening environment that has a live feel, which is easier for listening over extended periods. These goals should be discussed with the acoustician during construction.

Absorption Coefficients

The acoustic properties of absorption material are rated with *absorption coefficients*. Absorption coefficients indicate the frequency ranges where absorption is most effective, although absorption coefficients are not measured according to any standard. For example, absorption coefficients listed by one brand are not necessarily comparable with those of a different brand. While not exacting, absorption coefficients generally can help with finding the right material for addressing problem frequencies.

Diffusers

There are several types of diffusers. The main purpose of a diffuser is to scatter sound waves using an uneven surface. This is often accomplished in mastering studios with *quadratic residue diffusers* (QRDs). Some manufacturers go as far as to put miniature QRDs on the surfaces of their QRD wells. Diffusion also can be accomplished economically with something like thick, gathered vinyl curtains. There are also curved diffusers, grid diffusers, and many other types.

Front-Wall Treatments

There are two treatments that are common for the front wall behind the speakers: diffusion and absorption. Rooms may have a varying response to these kinds of treatments depending on the dimensions and construction materials of the room.

Treating First Reflection Points

One of the most important areas to treat is the *first reflection points*. These can be found by having someone move a small mirror along the wall and locating the places where the speakers can be seen from the listening position. These positions are often carefully treated with either absorption or diffusion, depending on the room's response.

Clouds and Ceilings

Acoustic absorbers, called *clouds*, are often used to treat the first reflection points on the ceiling above the listening position. Some studios use diffusers in the cloud position, although the best choice varies depending on the room. Ceilings are usually recommended to be at least 10 feet high at the lowest point.

Narrowing the Front of the Room

In many mastering studios, the room narrows toward the front. Each of the sidewalls, as well as the ceiling, has a slope. Nonparallel surfaces are thought to produce a more even frequency response by reducing standing waves and the effects of room modes.

Minimizing Noise

Noise should be minimized from air-conditioning systems, computers, and all other noise sources. In the United States, noise is usually measured by *noise criterion* (NC), a unit of measure used to specifically rate indoor noise levels. Outside the United States, a similar measurement called *noise rating* (NR) is used, as established by the International Organization for Standardization (ISO). These measurements can be made with a high-quality sound pressure-level (SPL) meter. Dolby Laboratories suggests a maximum noise level of NC 15, and any such noise should not be perceptibly impulsive, cyclic, or tonal in nature.

Noise also can come from sound vibrating an object or fixture; such sounds are called *sympathetic resonances*. Various tones can be played in the room to help locate these problems.

Rear Diffusers

One often sees diffusers on the rear walls of mastering studios and sometimes on the rear ceiling as well. While this is generally the best choice, as with nearly everything in acoustics, it may not be the best choice for every room.

Early Reflections

Early reflections are sounds that occur within a few milliseconds of the direct sound. The Dolby Theatrical Sound Production Facility Requirements define early reflections as ones that occur within 15 ms of the direct sound and suggest acoustic treatment so that these reflections are at least 10 dB below the level of the direct sound for all frequencies between 1 and 8 kHz. Absorbers and diffusers reduce early reflections. The Dolby specifications suggest that early reflections in the listening position should be at least 10 dB below the direct signal; others may recommend 15 to 20 dB below. This measurement is usually made with an impulse-response envelope, measurement microphone, and frequency analyzer.

Listen to Tones with an SPL Meter

If measurement equipment is not available, listen to different tones that are reproduced at the same level. Listen to how they sound in the room, and measure them with a sound pressure-level (SPL) meter to see which are exaggerated. This can provide some awareness of and insight into room problems.

Parallel Surfaces

Parallel walls and surfaces in a studio can create a surprising amount of reflections. Mastering studios typically have sloping walls and ceilings to address this. Even desk and rack surfaces can cause significant problems.

Console/Desks/Surfaces

Perhaps the most ignored source of acoustic problems is desks and consoles. Even computer monitors can cause acoustic problems. Mastering studios typically work to minimize flat surfaces, including console space.

Symmetry

While mastering rooms are normally rectangular and parallel walls are avoided, they are generally symmetrical otherwise.

Decoupling Monitors

Monitors are usually separated from their bases or the floor using spikes or foam. This helps to reduce the transference of energy from the monitors to their supporting surfaces and minimizes or eliminates sympathetic resonances.

Bass Traps

Bass traps are usually placed in corners if there are acoustic problems with bass reproduction. MondoTraps from Realtraps are a great option alongside many other bass-trap solutions on the market. There are also many do-it-yourself bass-trap designs available.

Minimal Objects in the Room

The furniture and objects inside the mastering room are minimized, especially objects with large, flat reflective surfaces. If they cannot be avoided, steps may be taken to make them less acoustically detracting.

Ideal Room Dimensions

There are several ideal room size ratios (between length, width, and height). When selecting a room or constructing a mastering studio, it should be fully researched. The very worst rooms are cube-shaped. Acousticians who have published ideal room ratios include L. W. Sepmeyer, M. M. Louden, Ben Kok, J. E. Volkmann, and C. P. Boner. The golden-rule ratio and the IEC-268-13, IEC 60268-13, Industrial Acoustics Company (IAC), and Dolby ratios also should be considered. The first listed on the following page by L. W. Sepmeyer and M. M. Louden are among the most popular room ratios. Typically, mastering rooms have at least 10-foot ceilings and a 12-foot width. Often, 30 feet of length is said to be most optimal because they allow deeper bass response, however this is untrue and many highly-regarded mastering rooms have much shorter depth.

A few of the most popular room ratios include

- L. W. Sepmeyer: 1:1.14:1.39
- L. W. Sepmeyer: 1:1.28:1.54
- L. W. Sepmeyer: 1:1.6:2.33
- M. M. Louden: 1:1.4:1.9
- M. M. Louden: 1:1.3:1.9
- M. M. Louden: 1:1.5:2.2
- J. E. Volkmann: 2:3:5
- C. P. Boner: 1:1.26:1.59 (rounded)
- Golden-rule ratio: 1.236:2:3.236
- IEC 268-13: 2.8:4.2:6.7
- IEC 60268-13: 2.7:5.3:7
- Dolby: 0.67:1:1.55
- ASHREA: 1:1.17:1.47
- ASHREA: 1:1.45:2.10
- Bolt: 1:1.28:1.54
- IAC: 1:1.26:1.60
- Kok: 1:1.12:1.41

Angled Walls Behind Monitors

In mastering studios, the monitors face the mastering engineer. In some highly regarded studios, a small angled wall is located behind the monitors in a way that is parallel with the face of the monitor.

Monitor Visibility

According to a study published in the *Journal of the Acoustical Society of America* (ASA), when monitors are visualized in a studio, it can affect one's perception of the sound. Therefore, some studios, such as Chicago Mastering Service, have their speakers behind acoustically transparent fabric where the monitors cannot be seen.

Correcting Acoustic Problems with Equalization

Most acoustic engineers agree that with most attempts to solve acoustic problems with equalization, more problems are created. Rarely, experienced acousticians are able to make certain fine adjustments using an equalizer, although this is virtually always for low bass frequencies only. Room acoustic problems are best addressed with acoustic design and treatments.

Speaker Placement

Speaker placement is of great importance in mastering. Often guidelines for placing speakers are based on the ITU 775 standard (International Telecommunications Union Operational Bulletin No. 775). ITU 775 specifies multichannel surround positioning for 5.1 surround-sound monitoring, which is shown in Figure 6-1.

However, many engineers feel that a slightly wider angle should be used for stereo monitoring than what is specified in ITU 775. Generally, for stereo monitoring, speakers are positioned forming an equilateral triangle with the listening position. This means that the three distances between each speaker and the listening position should be the same. As with all equilateral triangles, each angle is 60 degrees—that is 30 degrees to each speaker from the listening position

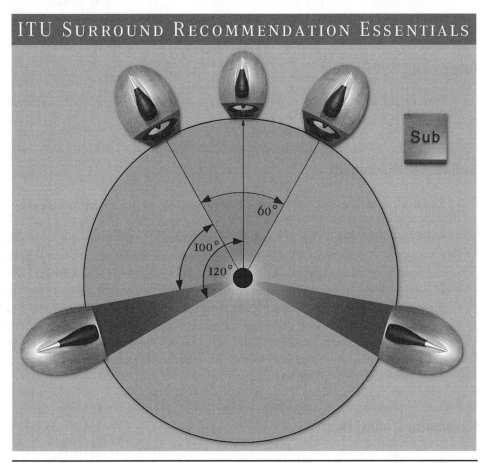

FIGURE 6-1   ITU 775 surround-sound monitoring standard. Many mastering engineers prefer a slightly wider angle for stereo monitoring.

if facing forward. To check the angles, some engineers position mirrors just above the tweeter and check to make sure that they can see themselves in the mirror at the listening position. Speakers should be placed at ear level. Although placement on one side of the room should mirror the other, the speakers are positioned so that the distance between a speaker and its two closest walls is not the same.

Also, because of what is called the *boundary effect*, the closer a speaker is to a wall, the more the lower-midrange and low frequencies will be exaggerated. Speakers often will be placed significantly away from walls unless the boundary effect is being used purposely. Also, speakers are not laid on their sides unless the manufacturer specifies them to be placed that way.

Speaker Decoupling

Speakers are usually placed on downward-facing spikes or foam as a buffer. This prevents the transfer of energy from the speaker to its base.

Integrating Subwoofers

If midfield or near-field monitors are being used, subwoofers are necessary in mastering.

Subwoofer Placement

Perhaps the best advice for subwoofer placement is to place them where they work best—each room is different. Generally, placing them too close to a wall can introduce problems. Also, it has been recommended that if a room were divided into a grid with four equidistant horizontal and vertical lines, subwoofers should not be placed on the lines or the intersection of such lines. These locations are said to be the worst for introducing room-mode issues. Also, subwoofers are usually placed either directly in line with the main monitors or slightly in front. Another method is to play music over the monitors that has plenty of bass, place the subwoofers at the listening position, walk around the half of the room where the main monitors are placed, find where the bass has the loudest response, and then place the subwoofers in those spots.

Subwoofer Calibration

Subwoofer calibration involves setting the subwoofer phase, crossover frequency, and level. The subwoofer *crossover point* is the audio frequency below which the subwoofer will reproduce sound. There are several methods for calibrating a subwoofer.

Method A

One of the most popular methods is to use a recording that features several bass notes across the spectrum that are at the same volume level, and adjust the crossover and subwoofer level until all are reproduced at the same level.

Method B

Some engineers use a recording of a bass sweep. The goal is to configure the crossover point and the subwoofer level so that the tones are reproduced uniformly across the sweep.

Method C

Play pink noise over the monitoring system, use a measurement microphone and spectrum analyzer to visualize the pink noise, and set the crossover frequency and level in such a way that the frequency response is most linear.

Method D

Dolby lists 120 Hz as the proper crossover point for theatrical film applications. For DVD and other consumer applications, a crossover point of 80 Hz is listed; in rare situations where the main monitors and the subwoofer have a crossover control, it can be set with this in mind if performing mastering for film.

There is an audio calibration tool that can help with these subwoofer calibration methods in Windows VST format provided freely on the Stonebridge Mastering website. It is called the Gebre Waddell Audio Calibration Tool.

Subwoofer Phase

Subwoofer phase should be set to the position that creates the highest loudness.

Excerpt from the Dolby 5.1-Channel Production Guidelines (the LFE Channel Is the Subwoofer)

"The LFE channel is calibrated such that each 1/3 octave band between 20 and 120 Hz is 10 dB higher than the equivalent 1/3 octave bands for any of the full-range speakers, assuming that the full-range speaker is ideally flat. This level is read from a real-time analyzer (RTA), rather than a sound pressure-level (SPL) meter. If an RTA is not available, an SPL meter may be used to approximate the level. When the level is correct, most meters will read around 90.91 dB SPL C-weighted slow for the LFE channel. The difference in level is because there is no energy being reproduced for the frequencies above 120 Hz (80 Hz for consumer applications)."

RT60/RT30/RT20

RT60 refers to the time required for reflections to decay 60 dBSPL below the direct sound from which they were produced. In some cases, RT30 or RT20 are measured, and RT60 is calculated from them. Although this is usually stated as a single number, it can be measured with a wide-band signal (20 Hz to 20 kHz) or in narrow bands (e.g., 1 octave, 1/3 octave, or 1/6 octave). This feature is built into many spectrum analyzers. RT60 times are often discussed in acoustics but are more relevant for large spaces, such as concert halls, than for mastering studios. With that said, RT60 targets for mastering rooms are usually 200 to 500 ms.

Listen to Translations

The term *translation* refers to how a recording sounds on various playback systems and in various acoustic environments. Acoustics problems in a mastering studio lead to translation problems outside the studio. The best remedy is to treat the acoustic options. If this is not feasible, engineers should be as aware of the problems as possible and work with them in mind. If masters from a studio are commonly translating with an issue, the engineer should begin paying attention to that issue. When starting off in mastering or with a new monitoring system or environment, it is important to listen on a variety of systems to develop a sense of the translation.

With some issues, the *reference level* can be adjusted, which is the loudness level of the monitors, while adjustments are being made by the mastering engineer. There is more on this after the next few sections.

Acoustic Methods/Techniques/Design Concepts

Several acoustic approaches may be used when designing control rooms or mastering rooms. The details are outside the scope of this book, although each approach is listed below for easy reference. The advantages and disadvantages should be researched and discussed with an acoustic engineer before planning.

- Front-to-back (FTB)
- Live-end–dead-end/reflection-free zone (LEDE/RFZ)

- Moulton
- Non-environment (NE), also called Hidley/Newell
- Ambechoic
- Controlled image design (CID)
- Reflection-rich zone (RRZ)

Selecting Monitors

Many engineers say that mastering monitors are best selected by auditioning them in the studio where they will be used. During this auditioning, very familiar recordings are played to gain better insight than a blind choice. Often professional engineers do not actually select monitors in this way, instead opting to make selections based on specifications or manufacturing techniques. This is likely due to the difficulty in arranging monitor demos. Overall, mastering monitors are selected that provide the best translation, imperceptible distortion, and precise detail. For more on this topic, check out "The Case for Full-Range Monitoring" by Scott Hull in Chapter 16.

High-Quality Amplifiers

A high-quality amplifier is as crucial to a monitoring system as high-quality monitors. Both are best selected in the same manner—by auditioning favored selections and making a choice based on one's preference. Mastering studios typically use the highest-quality class A or A/B amps, although some high-quality class D amps (from Hypex, Lipinski, Bel Canto, and others) are on the very same quality level. Class D amps have come a long way in quality and are often called *green* amps for their energy efficiency.

Biwire

Biwire is a connection between amplifiers and speakers. Some monitors support this feature, with each speaker having four terminals for making the connection (two negative and two positive). There are several stated benefits of biwire, although it has largely been rejected by audio engineers on the basis of the possible benefits being outweighed by the problems it introduces.

Background Information for Monitor Calibration

The information in this section is important on its own. However, it is presented here to provide necessary background information for calibrating a monitoring system.

Decibels   Decibels (dB) are often used to measure sound levels. Decibels are measured with a ratio instead of fixed units. Our ears can hear over a tremendous range of sound-pressure levels, making this kind of measurement necessary. For example, nearly total silence is 0 dB. A sound with 10 times the intensity is 10 dB. A sound with 100 times the intensity is 20 dB. A sound with 1,000 times the intensity is 30 dB, and so on. Intensity is separate from loudness and generally, 10 dB is twice the perceived loudness.

dBSPL   SPL stands for *sound-pressure level* and is used to measure physical loudness of sounds. SPL is a measure of atmospheric pressure in air. It is usually denoted in professional audio with dBSPL (e.g., 83 dBSPL). The sounds described above actually would be dBSPL.

dBFS   dBSPL is different from perhaps the most common decibel measurement in professional audio, dBFS (decibels relative to full scale). dBFS describes the level of a digital recording. 0 dBFS is the maximum digital level. The minimum dBFS depends on the bit rate; for 16 bits, the minimum is –96 dBFS; for 20 bits, the minimum is –120 dBFS; and for 24 bits, the minimum is –144 dBFS (each bit is 6 dB). dBSPL and dBFS are very different because a digital recording could be played back through speakers at any loudness depending on the amplifier output level. dBFS meters are digital meters, whereas dBSPL meters use a microphone to measure air pressure.

RMS   *Root mean square* (RMS), often referred to as an *average level*, is a measurement of the signal variance from zero. The average of a sine wave would always result in zero because the positive and negative values would average to zero. Therefore, an averaging calculation alone would not be meaningful to us. RMS performs squaring so all the values will be positive. Because the values are squared initially, the square root is taken to recover the value. The average is taken from these positive values to give us a meaningful result.

Calibrated Reference Levels/K-System   Bob Katz's K-System popularized the idea of monitoring at 83 dBSPL RMS in mastering and mixing. Simply stated, the idea is that when the RMS portion of the K-System meter shows a level of 0 dBFS RMS in the DAW, it correlates with the speakers playing at 83 dBSPL RMS in the room. This calibrates the monitoring system to a *reference level* so that the digital world (the meter) corresponds to the physical world (the SPL level). To accomplish this, one must adjust the output level of the amplifier/monitors. Usually this is done by playing pink noise at the intended RMS level (for the K-System, this is –20 dBFS for K-20, –14 dBFS for K-14, and –12 dBFS for K-12). Then, while using an SPL meter placed at the listening position (set to C-weighting, slow response), the amplifier output is set so that the SPL meter shows 83 dBSPL.

K-Meter Explained

A K-Meter has two parts, a peak meter and an RMS meter. It simultaneously displays both peak and RMS levels. There are three K-Meter types: K-20, K-14, and K-12. The main difference in these meters is how they are numbered on the side. The meters are marked so that the RMS that corresponds to their number shows as 0. For example, K-12 meters are marked so that –12 dBFS RMS shows as 0 on the meter.

K-20, K-14, and K-12

With Bob Katz's K-System, K-20 is to be used for film, K-14 for rock and other types of commercial music, and K-12 for broadcast and pop. Each of these meters is shown in Figure 6-2. According to the system, the compression, limiting, or even expansion should be set so that the peaks of the loudest passages of the recording reach the highest limit of the meter while the RMS level remains at 0.

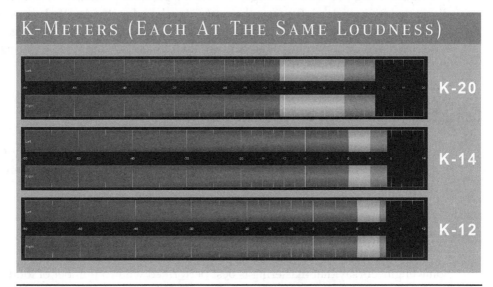

FIGURE 6-2   Bob Katz's K-Meters (K-20, K-14, and K-12) are intended for use in setting loudness during mastering: K-20 for film, K-14 for rock and other commercial music, and K-12 for broadcast and pop.

Stepped Monitor Gain Control

As discussed in Chapter 5, stepped controls click into a fixed number of positions across their range. Stepped monitor gain controls are important in mastering because they allow you to easily return to the exact level that you have calibrated. With a stepped gain control that is marked in decibel increments, it is possible to easily move between reference levels (e.g., K-12, K-14, and K-20).

K-System Criticisms

One of the main criticisms of the K-System is that 83 dBSPL is either too low or too high, which Bob Katz acknowledges. Also, while RMS metering is a closer representation of loudness than peak metering, it is by no means a precise representation of loudness. With regard to 83 dBSPL, Katz has suggested setting the level at your own personal taste. For some listening environments, this level may be too high; for others, it may be too low. The main idea is not the specific levels (K-12, K-14, and K-20); instead, it is working at a familiar reference level to develop a sense for how things should sound. While the K-System became popular among some engineers, by and large, established professional mastering engineers did not adopt it. Instead, they use their own developed methods. Working with a reference level is a technique that was in use before the K-System was created. Also, professional mastering engineers set the dynamics/loudness according to their ears and best judgment, not a meter value.

Consistency as a Result of Reference Levels

The main purpose of working with a reference level is to create consistency. Our frequency perception changes across loudness levels, so having a single reference level gives a sense for balance at that level.

Monitoring at a Variety of Loudness Levels

Some engineers say that mastering is best performed when a number of loudness levels are considered and monitored. Equal loudness contours show that our frequency perception changes depending on the monitoring level. Because mastering is about making things work in a variety of situations, it may be important to listen at different monitoring levels and make adjustments so that the sound works best across all levels of loudness.

Ear Sensitivity

Our ears have varying frequency sensitivity depending on loudness. The Fletcher-Munson curves (and their most recent evolution, the equal-loudness contours) describe how sensitive our ears are at various levels. The equal-loudness contours are shown in Figure 6-3. This is a factor in deciding a personal reference level. For example, if an engineer is producing masters that are too bright and have too much bass, he or she may consider raising the reference level, which changes the perceived frequency balance. Frequency perception based on ear sensitivity is one of the main factors for engineers developing their own personal reference levels.

FIGURE 6-3   Equal-loudness contours describing the sensitivity of the human ear across the audible frequency range. This graphic shows that the sensitivity changes depending on the loudness of the sound. We are most sensitive between 2 and 4 kHz, where the graph dips the lowest.

Alternative Reference-Level Selection

Some engineers set their reference level with the simple method of listening to recordings very familiar to them (or similar to the genre of the recording being worked on), finding the most comfortable level for listening, and using that as their reference level. Raising the loudness even 1 or 2 dB higher than this comfortable level may make a well-selected reference recording sound abrasive in the studio, although it truly is not. Therefore, if adjustments were made while mastering above the comfortable loudness level, we might attenuate harshness that would not serve translation.

There is also a practice of setting a reference level and adjusting it over time depending on customer feedback about translation. If masters are consistently dull and with too little bass, the level may be lowered. If they are too bright and the bass is just right, the reference level may be raised, and the subwoofer level may be lowered. There are many possibilities with this approach.

Workplace Safety

Most developed countries have some workplace safety standard for loudness. In the United States, the Occupational Safety and Health Administration (OSHA) specifies that a person can be exposed to 90 dBA for up to eight hours. This assumes that the recommended breaks are taken. Audiologists may recommend a maximum of 85 dBA.

Headphones

Headphones allow more detail to be heard, except in the bass range and the stereo field. While you can hear more detail, headphones distort linear frequency perception. Any adjustments that affect tone or that involve depth or wideness should be made with monitors. Mastering engineers generally use headphones only for quality control so that they can hear the finest detail. Headphone monitoring might be compared to a magnifying glass that has a distorted and discolored lens. More detail is perceptible, but the view is distorted.

CHAPTER 7

Mastering Practices: Techniques, Problems, and Approaches

In an interview with the Audio Engineering Society, Bob Ludwig described the "ah-ha" moment of his career. He discovered that a producer he worked with preferred the work of a different mastering engineer over his own. While Bob prided himself on his masters being as true to the mixes as possible, his competition would use compression, hype the midrange frequencies, make it brighter, and give a fuller-sounding bass drum. Bob's big realization was that the producer expected mastering to provide "more of the excitement felt in the mix room."

It has been quite some time since Bob's "ah-ha" moment while cutting vinyl, but many of today's producers are still looking for improvements of their mixes from mastering. This is especially so with home and project studio producers.

Truly improving a recording during mastering is quite a challenge and is often best done with subtlety. The main idea is to do what needs to be done while minimizing the impact on other elements of the recording. Sometimes preserving the existing tone is the best possible outcome. The word *tone* itself encompasses so many things. It relates to terms such as timbre, color, tonal quality, sound quality, and fidelity. Each one of these terms has its own nuances in our language. For the sake of simplicity, the umbrella term *tone* is used throughout this book.

This chapter presents a collection of techniques, approaches, and problems in audio mastering that affect tone. The purpose is to present a wide array of practices so that engineers can discover what works for them. These practices range from common to rare. Some help to make improvements, whereas others prevent degradation.

When techniques are described, they may come across as protocol, but it is the underlying principles that are most important. The better someone understands the principles, the more advanced and customized processing can become. This is not to undermine the techniques; they are certainly not irrelevant, but they should be considered a bit differently from instructions.

Technique: Listening

The most important skill in audio engineering is listening. Deeply listen to the song before and during processing. The song must be felt as much as analyzed. What does it feel like? What should it feel like? What can you do to help achieve the emotional goal—the story told by the song? In mixing, this experience is at its height. In mastering, the final refinement is taking place with a different perspective, experience, and tools. Mastering is about removing anything standing between the listener and the emotion created during production. If something can be improved, try it. If it is already perfect, that's fine. In such cases, mastering will be about minimizing artifacts while setting appropriate loudness levels and creating the destination format.

It is an important listening skill while mastering to focus on the adjustment being made and its effect on other aspects and pulling back to listen to the overall sound while considering the adjustment. Because almost everything in mastering affects more than one thing, benefits and sacrifices must be balanced.

Technique: Working with Intent and Vision of the Result

While listening, develop a vision for improvements—a goal to be reached. Lock onto the vision, and work to make adjustments that produce it. Developing this skill takes time and familiarity with techniques and the equipment at hand.

Technique: Only Making True Improvements

Adjustments should be compared at the same loudness level as the original mix to ensure that intended improvements are true improvements. Making true improvements is more of a challenge than it would first seem. Analog and digital processors alike can boost the loudness very slightly, even as little as 0.1 to 0.5 dB, which gives the impression of a subtle improvement. Such boosts sound like an improvement even when the processing has actually caused something undesirable. Because the ear almost always interprets slight increases in loudness as an improvement, the novice is tricked easily and often. Differences in level should never be allowed to mislead decision making about processing.

For a rigorous test of true improvement, set the processed output slightly below the level of the original, and compare the "improvement" with the original. This will provide insight into how much loudness affects the perception of an improvement. Typically, levels are matched as closely as possible during comparisons, which may be done by ear or with an LUFS meter.

There are several methods for making comparisons at the same loudness. The input offsets or DIM switch of a monitoring control system or console can be useful for this task. When working digitally, one might insert a plug-in with an adjustable gain control at the very end of the processing chain. It can be used to balance levels for comparison. A few tools for matching loudness are shown in Figure 7-1.

A similar but less influential auditory illusion is an effect caused by the visualization of some plug-ins. The graphically appealing nature of a graphical user interface (GUI) seems to activate the user's senses in the same way an audio improvement might. This effect can be experienced by watching and listening to a high-definition (HD) music video and then turning off the screen. The auditory experience can feel at least subtly different. In fact, there have been many studies that show links between our senses. To easily make the computer screen blank while listening, Windows users can make a shortcut to the "blank" screensaver on the Windows desktop (the file is usually C:\WINDOWS\system32\scrnsave.scr). Then right-click the shortcut, go into Properties, and set a shortcut key to quickly launch the blank screen whenever you want to focus on listening. The screen can also be turned off manually, although there can be a longer delay when powering on and off.

While developing mastering skills, it is of the utmost importance to make true improvements. Wasting time with illusory improvements takes away from time that could be used for actually developing one's senses and methods.

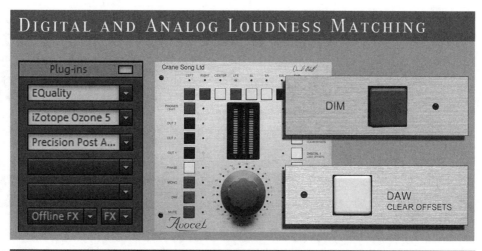

FIGURE 7-1 Various tools can be used to match loudness levels for accurate comparison, including plug-ins, DIM switches, and monitor-level offsets.

If making a true improvement does not seem possible with a given recording, that's okay. If this happens, do not feel bad about it. The only processing may be to raise it to suitable loudness with minimal artifacts and prepare it for its destination format.

This may be the most important point of this book: *Comparing any processing at different loudness levels is misleading and deserves careful attention.*

To help with comparing processing at the same loudness, a Windows virtual studio technology (VST) plug-in was created to coincide with the release of this book called the Gebre Waddell Precision Post Analog Controller. It features long-throw faders that allow for precise level matching and balance control. Holding the ALT key while moving the fader provides the most precise adjustments.

Approach: Destructive versus Nondestructive Processing

In audio engineering, there is destructive and nondestructive processing. This describes whether processing modifies the original content. *Destructive* processing changes the original content, and without a saved copy, the original content is lost. *Nondestructive* processing allows settings to be modified and saved in addition to preserving the original content. Nondestructive processing is familiar to audio engineers as the processing that takes place in a digital audio workstation (DAW) *session* view. Examples of session and waveform views are shown in Figure 7-2. Destructive processing is most familiar as the processing that takes place in a DAWs' *waveform* view. Most digital processing in mastering is performed nondestructively so that adjustments can be made later.

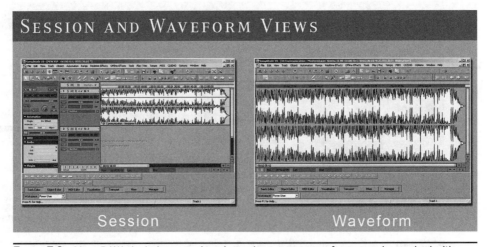

SESSION AND WAVEFORM VIEWS

Session   Waveform

FIGURE 7-2   Most DAWs include a session view, where many waveforms can be worked with together nondestructively, and a waveform view, where a single waveform can be worked with destructively.

Technique: Working with Reference Recordings

Reference recordings are usually popular recordings or recordings of the highest quality in a given genre and can be used for comparison alongside recordings being mastered. Engineers often maintain a collection of reference recordings and may request clients to submit reference recordings to gain insight about their tastes. When a client is trying to communicate his or her sonic vision, it is usually best accomplished through reference recordings. Also, working with a reference can help to maintain a sense of balance and help reset our ears and perception.

References are sometimes audio CDs or digital files. They can be loaded into the DAW or played on a separate device. Reference recordings are often routed to monitor control systems or console inputs so that comparisons can happen instantly at the touch of a button. This allows for instant comparison during both digital and analog processing. If these are not available, the DAW can be configured for the use of the "exclusive" solo mode as described in the next section. Whenever working with a reference, the same D/A converter should be used for both the original and the reference; otherwise, the comparison will not be exact.

Also, reference recordings should be stored in nonlossy formats. When reference recordings are stored on the DAW's hard drive, they may be ripped from CD to the Free Lossless Audio Codec (FLAC) format. This helps to minimize drive space and is useful for quick loading. It is usually accomplished using Exact Audio Copy (EAC) and the FLAC codec, both of which are free. With a Google search for both "exact audio copy rip to flac" and "eac flac command-line parameters," one can find a bit of help for properly setting this up.

Sometimes reference recordings are streamed from another device for playback, although you must be careful that reference recordings are converted with the same digital-to-analog (D/A) converter as the recording being mastered; otherwise, the comparison will not be accurate.

The annual Grammy Nominees compilation CDs are an inexpensive way of updating reference collections with essential popular music. Also, recordings from record labels focused on recording quality, such as Decca, Chesky, and Harmonia Mundi, are great for reference-quality recordings.

Technique: Minimizing the Delay Between Comparisons

Comparisons are made in a variety of situations—when working with reference recordings, between original and processed versions, or between several processed versions. Whatever the comparison may be, minimizing the delay between comparisons is important—seamless comparisons are absolutely essential to clear judgment.

Solo modes are a DAW feature that helps to make seamless comparisons. In mastering, the most popular solo mode is *exclusive*. When the solo mode is set to exclusive and a track's "Solo" button is pressed, all other tracks are unsoloed. This allows the user to switch seamlessly between tracks simply by soloing them. If the exclusive mode is not selected, groups of tracks will be soloed, and this does not work for making instant comparisons.

Mastering consoles and monitoring control systems usually have features that allow for instant comparisons. For example, the Crane Song Avocet allows input from many different sources, and the user can switch between inputs at the press of a button. This kind of feature allows for seamless comparisons between many types of sources. Of course, whenever digital sources are being compared, you must be careful that they are converted with the same digital-to-analog (D/A) converter and the same monitoring system so that you make a properly controlled comparison.

Technique: Avoiding Ear Fatigue

According to studies, there are two types of ear fatigue—short term and long term. Full recovery from short-term ear fatigue occurs in about two minutes. Because of this, mastering engineers may take a short break before committing to any final decisions. Recovery from long-term ear fatigue requires at least a few minutes but may take several days. Long-term ear fatigue can be induced from a loudness of 75 dBSPL and above for long durations.

Technique: Processing Sections of a Song Separately

Mastering largely involves finding settings that work over an entire song. However, when an engineer notices a problem isolated to a single part, he or she might decide to process that part separately or automate an adjustment.

Some mastering engineers have been known to automate some adjustments, such as stereo widening or extra compression during song choruses. However, this kind of processing is rare and largely viewed as extraneous. Automation of equalizer adjustments is more common, although still relatively rare.

In the case of removing a single click or pop, you might destructively edit a small piece, possibly saving a new version if the change potentially would need to be rolled back. These kinds of repair edits are often done in what many DAWs call the *waveform* view, which allows destructive editing.

If an entire passage has an issue, the section might be cut into a separate part and processed as if it's a separate recording. It is important to seamlessly edit the parts back together. When processing this way, phase shifts can cause problems in timing, especially when using processing that involves phase shifting, such as equalization. This may be noticeable as clicks, pops, or other anomalies at the beginning or end of an edit. Careful editing can help to address these problems, and automating the adjustment can eliminate the consideration altogether.

Technique: Minimizing Processing

The number of processors used, both digital and analog, are often minimized. Additional analog processors in the chain beyond what's needed can add unnecessary noise. Using several plug-ins in succession is especially seen as unfavorable. Minimizing the use of processing, especially with plug-ins, is a good practice.

Technique: Making Client-Requested Changes Properly

When a change is requested, the mastering chain is recalled, and the change is made with appropriate processing. It is important to fix the problem at its proper place in the sequence. It is valuable to save plug-in and analog processor settings to recall settings for making such changes.

Technique: Processing Based on First Impressions

Many engineers believe that their first impressions about what a mix needs are the most valuable. Adjustments are quickly made based on those first ideas. Working this way is a skill that takes time to develop.

While developing their skills, many engineers prepare a version, listen after a break with fresh ears, and make revisions as needed. This is meant to avoid the shifting perceptions and distorted judgment that can come from working with the same material for extended periods at a time.

Technique: Turning Things Off/Listening in Bypass Mode

Making true improvements requires bypassing processors to make comparisons. Processors may be bypassed individually and altogether. As mentioned earlier, making such comparisons at the same loudness is the only way to be sure that processing is making an improvement.

It may be best to make comparisons without knowing which is the original and which is processed, except by listening. There are various ways to accomplish such a comparison depending on the equipment being used. For example, using a button on a console instead of a toggle switch can aid in making blind comparisons. The button can be pressed a few times without thinking of how many times it's being pressed; then listening can occur without knowing whether the processing is engaged or not. Also, plug-in chain bypass switches can be pressed with the mouse button a few times until it is not known whether they are engaged or not; then the comparisons can be made.

It is important to remain indifferent with processing ideas, especially with unfamiliar processing. Sometimes they can be as rewarding to dismiss as they are to keep.

Technique: Relationships with Mixing Engineers and Producers

A strong working relationship with mixing engineers and producers can create an environment for the best processing and approach. These relationships certainly can help business as well. As with many types of relationships, communication and understanding are key in reaching mutual goals.

Technique: Concurrent Processing

Concurrent processing is processing with several or all processors working simultaneously (as opposed to processing in stages). Arranging a mastering chain for concurrent processing has several advantages. Processors can be adjusted to work in context with each other. This can help to minimize the processing that takes place and create the best combinations. Also, due to making adjustments in context, the strengths of each unit can be leveraged more fully.

Working entirely concurrently is simple when using only plug-ins or only outboard processors. It becomes more complex when using plug-ins both before and after analog processing. In this configuration, track input monitoring can be an option.

Sometimes groups of processors may be used concurrently while other groups are used in stages. For example, some engineers prefer to do all noise reduction in one stage and all other processing concurrently.

Technique: Processing in Stages

Processing may also be performed in stages. There are a few common trends with processing in stages. When restoration processing is required, many engineers perform it in its own stage, before other processing. Even when all other processing is done concurrently, this is still quite common.

Noncreative processing is well suited for a separate stage, such as DC-offset removal or sample-rate conversion.

If analog processing takes place, all analog processing is virtually always done in a single stage. This prevents multiple A/D and D/A conversions, which degrade the signal. Sometimes plug-ins are used concurrently with analog processing.

Some engineers prefer to work with limiters during the final stage and separate from previous processing. This is usually paired with an equalizer to correct the limiter's impact on tone. While this is convenient, many engineers do not find it to be ideal if concurrent processing is possible with a given set of equipment. This is because it introduces additional filtering.

When working in stages, file naming conventions and other organization become vital in maintaining recall possibilities and workflow.

Technique: Stem Mastering

Stem mastering, as described earlier in the book, is mastering performed from submixes called *stems*. When all the stems are played simultaneously (e.g., drums, guitars, vocals, and bass), they make up the entire mix.

There are two main ways that stem mastering is viewed. The first, and most common, is that mastering from stems forces the mastering engineer to have a different perspective with the processing. This negatively affects the final result or at least presents a greater challenge during processing.

The views on stem mastering have a wide range including extreme opposition and belief that it is suited only to problematic mixes. Some engineers see stem mastering with less opposition—they welcome the greater control but use it responsibly and with the mind-set of leaving the mix intact. Instead, the stems are used to minimize the common sacrifices in mastering. For example, de-essing may be used on just the vocal stem if it is overly sibilant, or bass frequencies could be cut or raised only on the stem where the bass instruments are present. Also, drum stems may be processed in a different way from the less transient stems.

No matter how stem mastering is viewed, it is almost universally thought to be a good option when a mixing environment is far less than ideal and when there are very significant problems that could not be addressed any other way.

When performing stem mastering, the mixbus compression used almost universally by professional pop and rock mixing engineers cannot be used with the stems. Because of this, stem mastering presents a drawback for genres where mixbus compression is typical.

Mastering is about the big picture. Mixing is about the individual tracks at hand. Stem mastering is not meant to replace mixing, instead it can allow for a different approach to mastering.

Technique: Reverb Processing

Reverb processing is taboo for most mastering engineers and rare for those who do use it. When it is used, it is almost always near the beginning of the chain, after any restoration processing.

A mastering engineer almost never would use reverb on a professional production. However, we do receive problem mixes at times that are in need of significant depth or spatialization. The first request would be to suggest a remix, but if that is not possible, adding subtle, high-quality, tasteful reverb may improve the problem.

The most basic approach is to use a reverb processor over the entire mix, perhaps coupled with a transient processor to regain punch that the reverb may reduce.

Whenever using reverb with any approach in mastering, the reverb predelay setting is usually carefully adjusted to achieve the best results. Reverb predelay delays the reverb signal, reducing the masking it might otherwise cause with the main signal. Also, some reverb units can be configured to work only with specific frequency ranges. With this option, usually reverb is applied only to the middle- and high-frequency ranges. Reverb on lower frequencies usually causes significant problems.

For an extremely subtle implementation, reverb may be added only to the side channel to add depth and to prevent affecting the punch of the primary instruments in the middle channel. With this approach, a transient processor usually is not needed because the punchy elements of the mix do not usually reside in the side channel.

Technique: Mastering Equalization

Equalization is responsible for most of the sound associated with professional mastering. In mastering, equalizers can be used to:

- Balance bass notes
- Remove resonances
- Address harshness
- Remove DC offset
- Highlight the vocal
- Highlight the best-sounding parts
- Add "air"
- Enhance tonality
- Minimize problems
- Enhance depth

The challenge is that these goals must be balanced. Often an improvement of one element requires the sacrifice of another. Engineers hone their equalizer balancing abilities over years and decades.

There are a number of different kinds of equalizers that one may encounter, including various analog and digital equalizers that may have several modes. Analog equalizers work by shifting the phase of an alternating-current (AC) signal and recombining it with the original signal to cancel certain frequencies. Digital equalizers work on the same principle except by using digital delays. It is worse for low frequencies due to pre-ringing artifacts. These phase shifts do not cause a *phasing* sound, although they can highlight existing comb filtering problems in a mix. Some digital equalizers have multiple modes that can be used, such as linear and minimum phase, each having a distinctly different character.

Parametric Equalizer Controls

There are three main controls on a parametric equalizer that control amplitude, center frequency, and bandwidth/Q. The amplitude of each band can be raised or lowered, the center frequency can be shifted, and the bandwidth, also called Q, can be widened or narrowed.

Analog Equalizers

Analog equalizers are preferred by most mastering engineers for all equalization tasks, especially at the far ends of the audible frequency spectrum. There are several reasons often given for why analog equalizers are thought to be superior to digital equalizers. These include the continuous nature of analog circuits, nonlinear distortion characteristics, unique phase-response characteristics, and accuracy problems with digital equalizers when trying to adjust high frequencies as they approach the Nyquist/sample-rate limit, even when oversampled. While the subjective sound of analog equalizers is widely preferred, there is little consensus about the reasons behind the preference. Weiss Engineering digital equalizers are the exception, which most engineers agree are of the highest quality and comparable with analog, although without providing the character of some analog equalizers.

Digital Equalizers: Minimum Phase

Of the digital equalizer types, minimum-phase equalizers are generally preferred for mastering tasks. These usually provide the most analog-like and musical sound of all digital equalizers.

Digital Equalizers: Linear Phase

Linear phase is generally not preferred in mastering. It is worse for low frequencies due to pre-ringing artifacts.

Technique: Using a Graphic Equalizer

Due to the compromised design concept of common graphic equalizers, they are seldom used in mastering studios for processing. Instead, they are typically used to control a compressor's side-chain signal, where filter quality does not affect the processing. There are rare exceptions such as the LC-based graphic EQ used for processing by Bernie Grundman.

Technique: Basic Frequency Balancing Using an Equalizer

Some engineers think of the frequency spectrum as having distinct ranges, with each having a unique nature. Typically, the spectrum is thought of in seven ranges: subsonic, bass, lower midrange, midrange, upper midrange, highs, and "air."

In the following sections, the symbol "~" will be used to mean "approximately."

Subsonic (~0 to ~25 Hz)

Let's start with the subsonic range. It is the lowest range and contains frequencies inaudible to the human ear. Because frequency content in this area takes comparatively enormous amounts of energy to re-create, it may be removed with a steep and transparent equalizer filter type, such as the Universal Audio Cambridge Equalizer's E6 or, an even better tool, the surgical Crane Song Ibis. Cutting these prevents speakers from working hard to produce frequencies we would not be able to hear.

Bass (~25 to ~120 Hz)

The bass range is different in nature from the ranges above it. It contains vastly more energy and has common issues particular to it, such as unintentionally uneven notes, resonances, and being relatively unbalanced compared with the ranges above it.

If bass notes are uneven and the producer agrees that this is not artistically desirable, a mastering engineer can use a transparent, surgical equalizer with a narrow Q value and a frequency chart to find the fundamental frequency of the weaker note and raise it to be even (or conversely, the stronger note can be lowered). If a resonance exists in the bass range, a notch or narrow Q parametric bell might be used to cut it. This might be done by sweeping the area with a medium Q, boosting and sweeping to find the offending frequency, narrowing the Q to shape it according to the resonance, and then cutting it (around –4 dB can be a good starting point, although it depends entirely on the recording). Such resonances are rare and have an out-of-place ringing sound when boosted, distinct from the resonance of the boost itself. Sometimes addressing resonances can create more of a problem than the resonance itself. It is important to verify that resonance corrections produce a true improvement.

Sometimes the entire bass range may be too high or too low and needs to be set. Often this is done carefully with a calibrated subwoofer in a well-treated acoustic environment and sometimes using a reference track to help keep things in perspective. Because there is not much else in this range other than the bass and bass drum, often this area is highly adjustable in mastering. The bass range has a nature of its own and can be especially pleasing to the ear. Don't let this mislead you, this range must be carefully set to a level that will translate well.

This might be accomplished with a roll-off, which is normally set higher, perhaps around 120 to 230 Hz and with a carefully set Q. Perhaps the best treatment is a low shelf that is generally set lower, perhaps around 80 Hz to 120 Hz, depending on the program material and equalizer at hand.

Lower Midrange (~120 to ~350 to 400 Hz)

The lower midrange area, like the others, has its own nature. Similar to the bass range, there is also a tremendous amount of frequency energy. Resonances are more likely to occur in this range than in the ranges above it. There are many instruments that have fundamental notes in this range that often overlap and cause such resonances. Given the high amount of energy in this range, addressing resonances can make an impact. Problems in this area can cause seemingly unrelated issues across the spectrum. The same procedure for addressing resonances described earlier for bass can be used in the lower-midrange area. Overall, resonances that require attention are not very common.

Addressing resonances in this area requires the highest quality of surgical equalizers. Even a decent-quality equalizer used for this task may not produce a true improvement. It is important to remain honest with judging such comparisons.

Some recordings may benefit from a bell-shaped boost in this area, perhaps around 300 Hz to add warmth to a recording that sounds thin.

Midrange (~350 to ~2,000 Hz/2 kHz)

There is significantly less energy in the midrange and higher than in the lower ranges. Resonances are not typically an issue in this range or the ranges above. This range might be raised or lowered with a bell shape to shape tonality.

Upper Midrange (~2 to ~8 kHz)

This area is of prime importance because it contains vocals. The vocal level may be adjusted if it sounds muffled or weak. If so, it may be boosted at a frequency that suits it best. Also, the human ear is most sensitive between 2 and 5 kHz, so using a bell curve with a center frequency in this range, perhaps at 3,150 Hz, can help to reduce harshness to the ear and help a recording fit with the equal-loudness contours.

Highs (~8 to ~12kHz)

This area may contain cymbal harshness, sibilance problems, or other similar issues, although harshness issues are usually more in the upper midrange. This is also an area that provides clarity, so it must be treated very carefully. Some engineers propose using a high-frequency roll-off to reduce harshness, whereas most mastering engineers disagree with such processing. For many types of music, a roll-off may produce a sound that is too dull. Instead, a shelf may be the best option to adjust harshness in this area or a wide or medium bell in the harsh-sounding frequency area.

Certain equalizers are overwhelmingly more suited than others for adjustments in this range as well as the "air" frequency range above it. In the era of vinyl, audio engineers wished that they could get crystal-clear highs and air. Now, for many styles such as rock and pop, a mastering engineer might work to maximize this area without being harsh. On the other hand, some blues, bluegrass, folk, and other styles may be best with a more warm/dull sound (less highs and "air").

"Air" (~12 kHz to the Limit of Hearing)

The "air" frequencies can provide a feeling of openness. Adding "air" to a recording that needs it can add a spatial sense to the mix. Some equalizers such as the Sontec and Avalon AD2077 are renowned for their ability to add "air." In the plug-in world, the Universal Audio Precision Equalizer is often seen as doing a decent job with this task. While these equalizers are commonly praised, there are others that also do a great job. For example, the Buzz Audio REQ-2.2 is great for adding "air" and almost any other task but is not usually mentioned as an "air" equalizer.

To add "air," most engineers will go to their favorite equalizer for this task, turn up the frequency on the highest band to its maximum, use a bell curve or a shelf depending on which sounds best with the unit and recording, and boost tastefully. If using a bell curve, sometimes the Q also will be adjusted to sound best for the unit and recording. Even when an equalizer's highest frequency is beyond the range of human hearing, this still usually works best. Baxandall curves are an option for providing a different flavor of "air." Some engineers use more than one equalizer simultaneously to add a complex "air" sound.

Out-of-Band Noise

Out-of-band noise, above the range of human hearing and above the Nyquist limit for a given sample rate, can cause an A/D converter to distort. This type of noise can be introduced when highs are boosted (especially when boosted significantly).

This can cause a harsh high-frequency sound. To prevent this, an extremely high roll-off may be used. The Dangerous Music BAX Equalizer offers such a roll-off at 70 kHz and other various levels. However, unless the high frequencies are of extremely high loudness, this is not a problem. For example, if a recording has 10-kHz peaking around 0 dBFS and the out-of-band frequencies are boosted in the analog domain, then the distortion from the out-of-band noise may cause a problem. With mastering quality converters and in almost all typical situations, this should not concern mastering engineers enough to do any processing.

Filters, Shelves, Bells, and Q Values

The typical parametric equalizer used for mastering is made up of several, independent filters. Each filter typically has its own gain, center frequency, and Q controls. Virtually all filters used in parametric equalizers are of a type called *second-order*, which is a reference to the amount of delay the filter uses to accomplish it's task. Many believe the ear functions in a way that is closest to this second order response.*

While filtering, we adjust the gain for the amount of filtering we want to do, we adjust the frequency center to the area that we want to affect, and the Q value for generally how wide we want to adjustment to span. A large part of the art of mastering resides within these adjustments. Wide Q values can be more subtle. As the Q narrows, it causes more phase distortion and larger group delay, which may have an audible effect depending on the amount of gain. The higher the cut or boost (at the same Q), the longer the *ringing*, a side effect of filtering. However, the narrowest Q values are thought to be surgical, allowing us to address problems such as resonances.

We use shelves to adjust entire ranges such as bass or highs. We use bell-shaped curves to adjust in specific areas. When harshness is present, a bell may be used at or around 3,150 Hz. We might also use a bell shaped curve in the lower midrange to add "warmth." The bass range is most typically addressed with a shelf. The highs may also be addressed with a shelf, while the air band may be addressed with a shelf or a bell, depending on the equalizer at hand.

To obtain the benefits of what's known as a *Butterworth filtering*, there has been a trend in mastering to use a Q-value of approximately .71 (rounded from $1 / \sqrt{2}$, or 0.70710678...). The benefit of Butterworth filtering is the minimization of frequency ripples in the *passband*. In other words, it has a more linear, flat response with regard to frequencies outside of those it is set to affect. When using any bell-shaped filter that is second-order, a Q-value of .71 does not create a Butterworth filter and there is no Q-value that would. With second-order filters, only high- and low-pass with a Q of .71 can be meaningfully called Butterworth and obtain the

---

* However, higher-order filters, such as fourth order, with practical usefulness perhaps up to sixth order, may produce flatter passbands and sharper bandedges, as shown by Dr. Sophocles Orfanidis in his paper "High-Order Digital Parametric Equalizer Design." Bell-shaped Butterworths and other such filters are possible at higher filter orders, unlike with second-order filters.

benefit of a maximally-flat passband. To summarize, Butterworth is something that is only possible with high- and low-pass filters, not bell-shaped filters, with virtually all mastering equalizers.

There are other filtering methods with benefits, such as *Bessel* (maximally flat group delay with maximally linear phase within the passband; Q value of 0.58 rounded from 1 / √3, or 0.57735026...), *Chebyshev* type 1 and 2 (similar to Butterworth but with steeper roll-off and thus more ripple), *Elliptic* (also known as *Cauer*, equal ripple in stopband and passband and steepest possible roll-off) and Papoulis (compromise between Butterworth and Chebyschev), none of which can be accomplished with bell-shaped second-order filters.

For more on equalization filters, see Pieter Stenekes' "Digital Filtering" in Chapter 16.

Final Word

As mentioned previously, techniques should not be taken as protocol but rather as a guiding approach that contains insight. The advice in this section describes a single approach that can vary depending on the engineer, genre, and recording at hand. Frequency range designations become less important in the upper-midrange frequencies and higher. There are engineers who think of equalization as an entirely tunable experience and do not consider the frequency spectrum in separate ranges.

Technique: Order of Frequency Adjustment

Some engineers prefer to work with bass frequencies first and work up from there. Owing to the energy needed to reproduce them, lower frequencies have the greatest impact on overall sound. Other engineers prefer to work in a less organized fashion, which is likely the most common approach.

Also, surgical changes typically precede the more broad, tone-shaping changes. Finally, if an issue cannot be resolved with equalization, other methods may be used, such as side-chain compression, multiband compression, or dynamic equalization.

Technique: Substractive Equalization

Subtractive equalization is a technique where frequencies are only cut instead of boosted. Cutting frequencies in one range can result in the same or similar effect as boosting another range. Because any boost in loudness can be perceived as an improvement, if something seems better with a cut, it is most likely a true improvement. Also, some equalizers function in such a way that cuts produce a better sound than boosts. Few mastering engineers subscribe to this method in a strict way, although the idea behind it is relevant and considerable.

Technique: Using Less Common Equalizer Filters

There are a few less common equalizer shelves that some mastering engineers find useful.

Baxandall Shelves

The Dangerous Music Bax and some software equalizers offer *Baxandall shelves*. The circuit design that originally produced these curves was invented by Peter Baxandall in the 1950s. They are essentially shelves with such a wide Q that they often do not level off within the audible range. Baxandall shelves have a *shoulder frequency*—the frequency at which they level off into a horizontal shelf. Working with a shoulder frequency of as low as 40 Hz may affect such a far-away frequency as 4 kHz. Baxandall curves are mostly used in the same manner as a shelf, except with more transparency, and they are very popular for adding "air." A representation of Baxandall curves is shown in Figure 7-3.

Gerzon Shelves

Michael Gerzon was a mathematician and audio engineer who, among many things, created *Gerzon shelves* as a way to emulate the action of classic equalizers. The shape of a Gerzon shelf can be seen in Figure 7-3.

Niveau/Tilt Filter

Niveau and tilt filters, such as the Elysia Niveau Filter or the Softube Tonelux Tilt, work in a way that tilts the frequency response. With these equalizers, there is a configurable center frequency that allows the axis of the tilt to be set. An example of the frequency response of these filters is shown in Figure 7-3.

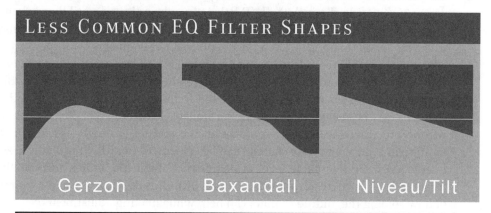

FIGURE 7-3 Gerzon, Baxandall, and Niveau/Tilt shelves are less common but can be found in use by some mastering engineers.

Technique: Frequency Roll-Off on Both Ends

Some engineers recommend roll-offs on the extreme ends of the frequency spectrum using an equalizer. While this seems to be a popular recommendation, it is not nearly as common among professional engineers as it may seem.

There are a vast many engineers who may perform roll-offs on the subsonic frequencies but never the high frequencies. Instead of roll-offs, shelves also may be preferred on the lower and subsonic frequencies. Of course, the choice of any such treatment depends on the recording at hand.

The purpose of high-frequency roll-offs is usually to address harshness. A later section provides many more options for dealing with harshness.

Technique: Extra Equalizer After Compression

Sometimes using an equalizer after the compression in the chain can yield a good result. Even a simple one-band shelving equalizer can prove useful in creating a frequency balance. In the analog domain, these are sometimes used after the compressor and before the conversion to digital.

Technique: Monitoring the Middle and Side Channels

Soloing the middle and side channels of the mix at hand, along with those of reference recordings, can provide useful information. Also, listening to the side channel can help with correcting left/right (L/R) balances to check for drifting of the stereo image, especially when analog processing has taken place. If instruments that should be in the middle channel (e.g., snare, vocal, bass drum, and bass) are present in the side channel, the L/R balance can be adjusted until those elements disappear or become as minimized as possible. This can be quite a useful technique.

Even when making L/R adjustments, some engineers will perform the adjustments with the middle channel soloed. This allows *zooming in* to the main elements of the recording. For example, de-essing might be performed with the middle channel soloed while adjusting the settings. Also, both middle and side channels may be soloed to find offending sibilance, and only the offending channel will be processed (usually it would be in the middle channel). When applying limiting in L/R, the middle channel can be soloed afterward in the chain to zoom in and judge an acceptable level of degradation with the limiter and exactingly locate the point where audible distortion is introduced. These are a few examples of how soloing the middle channel can help while making adjustments.

Technique: Mid-Side Processing

When an issue exists only in the middle channel, some engineers will make corrections only in the middle channel.

A few engineers find that dynamics processing/compression is best suited for middle-channel processing only because the side channel is rarely overly dynamic. This approach has not been widely adopted. Most engineers dislike any mid-side processing (other than filtering) because of how it disturbs the stereo image.

There are also a few who combine multiband compression and mid-side processing to work toward transparent control when there are specific dynamics problems to address. Perhaps the most common application of this is with middle-channel de-essing.

Another mid-side equalization technique is to embellish the high and lower midrange frequencies of the side channel while enhancing the midrange frequencies of the middle channel. There are a few engineers who adjust the lower frequencies of the upper midrange (somewhere between 350 and 650 Hz) with a slight cut on the sides and a corresponding boost in the middle.

Also, a few engineers boost with a low shelf on the side channel starting around 600 to 650 Hz, setting it to taste. This was discussed in Michael Gerzon's "Stereo Shuffling: New Approach, Old Technique." However, most engineers wish to avoid low bass in the side channel.

When working in mid-side, any processing that takes place on the middle or the side will affect the other channel to some degree. Because of this, adjustments made to the mid or side channels are done with this in mind.

Although one may encounter these techniques discussed as being widely applicable, as with any adjustment in mastering, all adjustments depend on the recording at hand. There are several well-respected mastering engineers who hardly, if ever, use any mid-side processing, although there are equally respected engineers who use it regularly, with their own developed techniques. To find out how one seasoned mastering engineer uses mid-side processing, see Brad Blackwood's "Mid-Side Processing" in Chapter 16.

Technique: Checking Mono Compatibility

While working in mid-side, some engineers recommend checking *mono compatibility*, which entails listening to how a recording sounds when converted to mono. Other engineers do not worry about mono conversion and see it as irrelevant—they would not sacrifice anything with their stereo image for mono compatibility.

It has been said that listening with a car stereo is more like listening in mono owing to the arrangement of the speakers. Also, when FM radio stations become weak, some receivers will switch into mono mode, and many AM stations today broadcast in mono. These are a few reasons why some engineers believe that it is worth it to give the masters a listen in mono.

Many mastering consoles, monitoring control systems, and DAWs have an easy-to-use feature for listening in mono.

Technique: Using Unique Mid-Side Processors

Basic mid-side processing allows for the middle or side channels to be raised or lowered. With analog processing, there are several mid-side converters that can be used so that any analog device can be used in mid-side mode. Also, some analog processors have mid-side functionality built in.

Of course, there are digital plug-ins that help with mid-side processing as well. This section describes a few of the unique digital mid-side plug-ins.

DDMF Metaplug-in/Mid-Side Plug-in

With a combination of the DDMF Mid-Side plug-in and the DDMF Metaplug-in, any VST can be used in mid-side mode.

iZotope Ozone

The iZotope's Ozone bundle offers a range of effects that all can be operated in mid-side mode.

Brainworx

Brainworx has focused on creative mid-side plug-in processing. This is reflected in both its bx_digital equalizer and the XL.

Mathew Lane's DrMS

This plug-in has unique "focus" and "field" controls that allow middle information to be sent to the sides and side information to be sent to the middle channel. This can be especially useful when working with stems.

Technique: Understanding Distortion/Coloration/Saturation

When discussing this subject, one might hear the terms *coloration, distortion, linear distortion, nonlinear distortion, intermodulation distortion, harmonic distortion, even-*

order/odd-order distortion, and *saturation*. *Coloration* is the most general of all such terms. It describes any change in the sound that the human ear can detect. *Distortion* is another general term, referring to any of the more particular types of distortion.

Nonlinear distortion describes a nonlinear relationship between the input and output signals, with a frequency component not present in the original signal. There are two types of nonlinear distortion: harmonic and intermodulation distortion. *Harmonic distortion* refers to distortion that occurs at a multiple of the original signal. *Intermodulation distortion* does not occur at multiples of the original signal.

Harmonic distortion, which occurs at multiples of the original signal, has two types: *even-order distortion* and *odd-order distortion*. These refer to whether the multiple is an even multiple (e.g., 2, 4) or an odd multiple (e.g., 3, 5).

Saturation is a term that describes a combination of distortion and compression.

Linear distortion is not a term used very often in mastering. It describes a distortion of amplitude or time (phase).

In mastering, distortion is sought that is pleasing to the ear. In the analog world, certain types and configurations of vacuum tubes, tape, and transformers are known to create pleasing distortion. Digital algorithm designers have drawn on the discoveries of analog engineers to create emulations and new processes based on them. Read more from one of the leaders of harmonics processing, in Dave Hill's "Distortions and Coloring" in Chapter 16.

Technique: Using Digital Emulations of Classic Gear

Some of the faithful emulator plug-in designers such as Universal Audio, Waves, and Slate Digital have products that emulate transformer distortion and other nonlinear characteristics from classic equipment. There are some pieces where just "running it through" adds the unit's characteristic distortion. This is meant to simulate the tubes, tape, or transformers of the analog domain. A few examples include the Universal Audio Pultec and Massive Passive, Slate Digital's Virtual Classic Console and Virtual Tape, and Waves PuigTec EQP-1A.

Technique: Running Through Twice

With some types of music where adding a high amount of distortion may be desirable, a mastering engineer may wish to run through an analog processor more than once or insert more than one instance of a digital processor that creates distortion. Of course, when experimenting with a process such as this, one must be very mindful to ensure that true improvements are being made.

Technique: Adding Distortion to the Side Channel

Adding distortion to the side channel is sometimes better for adding character than applying it to the entire mix. Because the ear wants to hear voices clearly, too much distortion on vocals can be disturbing. The Brainworx XL plug-in was made with this technique in mind.

Technique: Using the Same Character Processors on All Songs

Processors that impart a particular character may be used on all songs of a collection to give them continuity and a related sound, making the songs sound like they belong with each other.

Technique: Using Dither

Dither is noise added into the signal, noise that only subtly affects the sound of a recording. The purpose of dithering is to remove the digital errors that result from reducing the bit rate.

Dithering in the Visual Realm

Dithering is a concept used in both audio and visual media. The graphic in Figure 7-4 shows how various visual dithering techniques affect a gradient/fade when

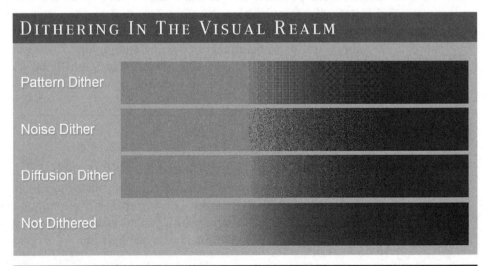

FIGURE 7-4 How dithering looks in the digital imaging. Although this is quite different from professional audio, it helps to visualize the concept.

the number of shades is reduced. This helps to visualize and understand the concept of dithering.

Audition Different Types

Dithering algorithms can be auditioned to find which fits best for the song. Currently, MBIT dithering set to *ultra* mode is one of the most popular.

Impact of Dither on the Sound

While the tone of dithering algorithms can sound slightly different, it's just not that big of a deal.

Dithering Should Be Kept to a Minimum

Because dither is essentially adding noise into the signal, it is something that should be kept to a minimum. It should be done whenever the word length is reduced (e.g., 24-bit to 16-bit conversion). Often, this happens only once. Typically, it is used at the very end of the processing chain. Often dithering is built into digital limiting plug-ins, which are found at the end of the chain.

Dithering for Digital Processing

Some digital processors work at a maximum of 24 bits. Because of this, when the source is 32-bit float, it should be dithered and converted to 24 bits if a processor has a 24-bit maximum. Otherwise, bit truncation could occur. Some mastering engineers work exclusively at 24 bits to avoid this possibility.

Problem: Jitter

Jitter is somewhat of a general term describing anomalies with digital signal transmission. Jitter may have a glitch or static-like sound and even may produce dropouts at its worst. It also can produce subtle effects such as reducing the sense of depth, focus, and clarity. In mastering, jitter is mostly avoided with high-quality wordclocks, correctly distributed wordclock signals, and a high-quality audio interface that minimizes jitter. Mastering studios often use AES/EBU connections along with BNC wordclock connections to distribute clock signals between digital devices because this configuration produces the least jitter. High-quality A/D and D/A converters, short cable runs, and correct cable impedances also keep jitter to a minimum.

Problem: DC Offset/Asymmetrical Waveforms

Today, problems with DC offset are quite rare, although possible. DC offset is when the entire waveform is shifted up or down, which is distinct from the more common *asymmetrical waveforms*. DC-offset problems are usually caused by a wiring, equipment, or microphone problem and can cause loudspeakers to work inefficiently. Asymmetrical waveforms are not necessarily a problem and can be caused naturally by several sources, including brass instruments. Mastering engineers might address either of these with a steep high-pass filter at a very low frequency before any other processing is done. This kind of high pass also might be used to solve problems with infrasonic rumble, which is very low bass noises that can be caused by air-conditioning systems, airplanes, and other ultra-low-frequency noises during recording. An example of an asymmetrical wave is shown in Figure 7-5.

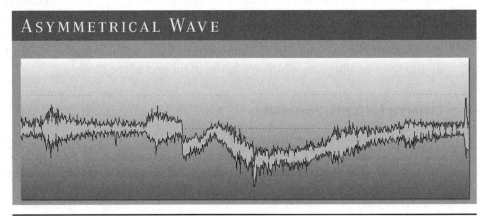

ASYMMETRICAL WAVE

FIGURE 7-5 An asymmetrical wave is not necessarily a problem; it can be caused by several sources, including brass instruments and some synthesizers.

Technique: Bass Enhancement

Harmonic and resonance processing can be used for bass enhancement. However, many mastering engineers only use equalizers for this task. Two of the most widely recognized harmonic bass processors in mastering are the SPL Vitalizer (hardware and plug-in) and Waves MaxxBass. These processors add harmonic distortion related to the bass frequencies.

Resonance processors, such as the Little Labs VOG, are based on special resonant equalization curves to create a full-sounding bass effect. Most mastering engineers prefer to obtain this kind of effect from a mastering equalizer, although the use of this kind of processing is not completely unheard of in mastering.

Problem: Bad Mixes

Major problems are usually repaired before other processing. Most engineers address serious problems digitally, before any other processing occurs. This includes major frequency imbalance, noise problems, and so on.

When a mix has a major frequency imbalance, multiband compression can help to even things out. The way radio stations operate is worth considering. The Orban and Omnia processors are very widely used in radio, and one of the primary processes of these units is their multiband processing. These processors make nearly everything sound even and can be thought of as a kind of automatic mastering. In today's world, an experienced mastering engineer will always be able to create a better sound than any such automatic method, but you can take from this that multiband compression is a treatment that can work when there are major problems to make them sound acceptable. Of course, every mix is different, so the multiband compression should be set according to the mix.

Spectral dynamics processors, such as Voxengo's Soniformer, provide another option when working with a mix that has significant problems. Soniformer separates a mix into 32 spectral bands and allows dynamic control over each band. Setting it up can be done quickly and can provide transparent and musical processing.

Multiband processing is not nearly as common in mastering as one might believe from the way the processors are marketed. It's worth keeping in mind that the use of multiband processing (other than de-essing) is rare on professionally produced recordings, and many very well-respected mastering engineers almost never use this kind of processing.

Problem: Hum in the Analog Signal Chain

When trying to find hum in an analog chain, it is usually best to disconnect devices and cables one at a time to isolate the problem. If a device has a problem with hum, there are two main options. First, if it has a grounding screw, a thick wire can be connected from it to some central (often large) metallic surface in the studio. Examples of the symbols that usually indicate the location of a grounding screw are shown in Figure 7-6. The second option should be done at your own risk, and that is to remove the ground part of the balanced cable connecting that device, which can eliminate the hum but can possibly cause more hum or damage the device.

WARNING *Although unlikely, removing the ground from a balanced cable can damage a device. It may be best to consult with a technician when working to address hum.*

If there is hum that is coming from the amplifier, even when there is nothing connected to the amplifier inputs, this can indicate issues with the amplifier's

GROUNDING SYMBOLS

FIGURE 7-6 Two common symbols for grounding that can be found on many pieces of equipment.

capacitors or possibly other components. Replacement capacitors and other parts usually can be ordered from the manufacturer.

If the services of a technician must be sought, it is best to provide him or her with notes of the problem and details of any attempts to fix it.

Problem: Sibilance

Sibilance is the "*s*" sound of vocals that's sometimes sharp and strident. High sibilance is a somewhat common problem with mixes. Sibilance is best addressed in mixing, but often mastering engineers are faced with mixes that contain too much. Processors that help to reduce sibilance are called *de-essers*. De-essers also can help with harsh cymbal sounds and other harsh high-frequency sounds. There are many mastering engineers who perform de-essing in mid-side mode, only on the middle channel, because that is where vocals most commonly reside. In some situations, sibilance during tracking causes actual distortions in the high frequencies, which cannot be repaired with de-essing.

Technique: Raising Levels Before Analog Processing

Some engineers compress, normalize or otherwise raise levels digitally before analog processing. The noise that can be introduced by A/D and D/A converters is minimized when taking this approach. All A/D and D/A converters have a noise floor. It is important to perform this technique in stereo-linked mode or somehow ensure that level changes are exactly the same on both the left and right channels to avoid any problems with unintentionally modifying the left/right balance. Analog equipment can sound better at lower levels, so this technique must be implemented with care.

Technique: Adding Noise in the "Air" Band for Brilliance

Adding subtle noise/distortion in the highest band, approximately between 16 and 22 kHz, can add what many perceive to be *brilliance*. There are several tools that can help with this, including iZotope Ozone.

Problem: Limiting Distortion

When limiting is too high, it can cause distortion and a harsh sound that causes ear fatigue. It also can cause consumer D/A converters to distort severely even though a mastering D/A converter may not.

Problem: Intersample Peaks

To understand intersample peaks (ISPs), you must first understand reconstruction. With a digital signal, the highest possible level is 0 dBFS. The digital signal must be converted to an analog signal for listening. Generally, this is called *digital-to-analog (D/A) conversion*, but the method of the conversion is called *reconstruction*. In reconstruction, if there are two simultaneous 0-dBFS samples, the actual reconstructed analog level may go above the level that corresponds to 0 dBFS, which is described as an *intersample peak*—a peak that occurs between digital samples. It exceeds and does not precisely correspond to the samples. Figure 7-7 may help you to understand this idea.

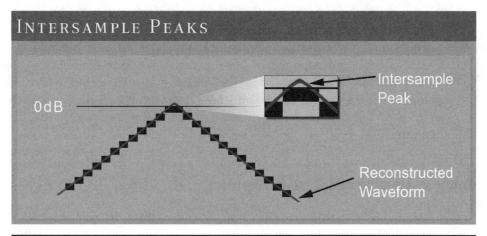

Figure 7-7   When audio is converted from digital to analog, intersample peaks can go above the highest intended level. There are techniques and tools to address this possibility.

To avoid intersample peaks and other problems with consumer D/A converters and format conversions, many engineers use a limiter ceiling of –0.3 dBFS.

Technique: Using Sample-Rate Conversion

Sample-rate conversion is entirely a digital process. Whether hardware or software is used, the only thing that matters is the algorithm. In fact, according to some tests, there are software sample-rate converters that outperform hardware.

Many mastering engineers prefer to use sample-rate conversion after limiting and before dithering. In this way, all processing can be performed at the higher sample rate, which may provide a benefit when processing digitally.

When sample-rate conversion occurs, intersample peaks may be a side effect. Because of this, some engineers perform sample-rate conversion in sequence immediately before the final limiting (with some type of intersample-peak prevention). The purpose of this is to gain the benefit of working with digital processors at a higher sample rate but to avoid intersample peaks.

Problem: Using Unbalanced-to-Balanced Connections

Using a combination of balanced and unbalanced connections is not necessarily a problem. The Pendulum Audio compressors, which are very popular in mastering, use unbalanced XLR connections and are typically used with balanced gear.

Problem: Unbalanced Left and Right Channels

Some engineers say that the balance between the left and right channels simply should be adjusted to where it sounds best, the most cohesive, or *locked*. There are a few producers who are very particular about any adjustment being made to the stereo field during mastering. For these clients, it should be carefully preserved.

There is also a technique, described in the mid-side section, where the side channel is soloed, and the L/R balance is adjusted until the typical middle-channel sounds (e.g., vocals, snare, bass, and bass drum) are completely inaudible, or minimized.

Problem: Unbalanced Bass in the Mix

In mixing, bass instruments (e.g., bass guitar and bass drum) are almost always centered. There are some rare recordings where they will be intentionally mixed more to the left or right side.

If the bass is mixed only very slightly to one side or the other, it usually indicates an intention to center it. In mastering, we may help it along to become centered. Soloing the side channel can reveal whether the bass is panned—if it exists in that channel, it has been somehow panned.

In the vast majority of mixes, bass instruments such as bass guitars and bass drums are centered. Bass frequencies require the highest amount of energy to re-create. Using both speakers allows for the most efficient reproduction of the bass frequency energy. Also, bass frequencies are, for the most part, nondirectional. This means that we do not hear as much, if any, stereo effect with them. For these reasons, bass frequencies are virtually always centered.

With the weight that bass frequencies carry, a mix can feel lopsided if bass is panned. However, there are recordings that work well with panned bass guitar as an intentional effect. This is especially popular with some jazz recordings. However, even with these albums, often the bass frequencies will be centered in mastering, or before mastering.

The term *elliptical equalizer* is typically used in vinyl mastering. Elliptical equalizers or equalizers with a similar function are used to convert bass frequencies from stereo to mono. The Brainworx bx_digital Equalizer has a simple function called Mono-Maker for converting bass frequencies from stereo to mono below a configurable frequency. Also, the Matthew Lane DrMS plug-in offers this kind of functionality with the highest configurability. Another option is to use a high-pass filter on the side channel with an equalizer that has a mid-side mode or that is used with a mid-side converter.

When bass is not centered, there can be problems with mastering the recording to vinyl at high levels, although cutting at lower levels can make it work, along with a few other methods.

Technique: Upsampling

Upsampling is the practice of using a sample-rate converter to convert a recording of a lower sample rate to a higher sample rate. In mastering, there is a technique of upsampling before performing digital processing to improve the quality of the processing.

While many engineers agree that upsampling can improve the sound of digital processing, it is also possible to degrade the signal with upsampling. This is especially possible with extreme upsampling at levels as much as eight times or more of the original rate.

Today, many digital processors have high-quality upsampling built in, running at a higher rate in the background so that it does not have to be done manually. Upsampling is a very popular feature of modern digital limiters because it can help with the reduction of intersample peaks. Because of the upsampling built into most of today's digital processors, it is less important to perform manually.

Technique: Digital Limiting

Digital limiting typically is used near the end of the processing chain, just before sample-rate conversion, dithering, and conversion from 24- to 16-bit format. Limiting can have a dramatic impact on the dynamic range and sound quality of a recording. Carefully comparing a processed version with the original can reveal differences. Such comparison is usually done manually, although a simple-to-use comparison function is built into the Slate Digital FG-X limiter with its "constant-gain-monitoring" feature. Without such a feature, comparison is still easy; it can be accomplished by inserting a plug-in with a gain control after the limiter and adjusting it until the original and the limited version are at the same level. The entire plug-in chain can be enabled or bypassed to compare with the original.

In mastering, it is most common to use a limiter output ceiling of –0.3 dBFS when limiting. This prevents problems with consumer D/A converters and format conversions. Most limiters have an output-ceiling option, although setting the ceiling also can be accomplished by lowering a track fader, master fader, or object fader in the DAW software by –0.3 dBFS with the limiter at a ceiling of 0 dBFS.

There are multiband digital limiters, such as the Waves L3, that are preferred by some engineers. There is somewhat of a trend of using these for dance music. However, most engineers do not prefer this type of multiband digital limiting for any task or any genre.

Typically, when applying limiting, special attention is paid to the loudest passages because they will be affected the most by limiting. Sometimes the middle channel is soloed in the chain, after the limiter, to listen for distortion effects more precisely.

Problem: Lack of Vocal Clarity

When the vocals sound muffled or in the background, the most common treatment is to do a subtle boost in the area around 2.5 to 4kHz or a cut in the lower midrange with a bell curve. This equalizer adjustment might be tried in the middle channel. Sometimes adding "air" is the best treatment.

Problem: Matching Equalizers

In the misled spirit of working according to visual references, some novice engineers use matching equalizers such as Har-Bal and the matching features of

Voxengo's Curve EQ. This type of processing is based on false assumptions. Mastering with a matching equalizer according to a reference track will not provide the effect of imparting the same or similar tone. This approach is borne out of the desire for an easy solution to a complex task—a task that simply requires experience and skill. Working to match a spectrum/fast Fourier transform (FFT) display of a reference track while mastering would be as absurd as working to match such a display while writing a song or recording. There is much more to mastering equalization, recording, and song writing than a frequency-response display.

Problem: Harsh/"Digital" Sound

Harshness is often very best addressed with careful bell-shaped cuts, most commonly around 3,150, 2,700, or 5,000 Hz. This area is where the ear is most sensitive.

Addressing Resonances

It is within the realm of possibility that a harsh tone in the high frequencies can come from lower-midrange resonances. A technique for addressing resonances was described earlier in this book.

Raising the Lower Midrange or Bass

Raising the lower-midrange or bass region with a bell curve or a shelf can help to balance a mix if the harshness is caused by excessive highs.

Rolling Off the Highs

One might find recommendations to roll off the high frequencies, although this can result in a very dead sound. Most often the very highest frequencies, which are affected the most by a roll-off, are not the source of harshness. Because of this, many engineers would virtually never consider rolling off the high frequencies.

Using a High Shelf

A high shelf can be a better treatment than a roll-off if a wide range of frequencies must be affected.

Dynamic Equalization on Upper Midrange/Refinement

Dynamic equalization can help to transparently reduce harshness without having a noticeable effect on other elements of the mix. The Weiss EQ1-DYN is perhaps the most popular for this treatment, with some engineers finding it indispensable, especially in today's world of harsh digital mixes. Also, I developed the "Refinement" plug-in while writing this book to be the best way of precisely addressing harshness using dynamic filtering.

Using a De-Esser

High-quality de-essers, which are basically dynamic equalizers, are sometimes used to reduce high-frequency harshness. One of the most popular is the Maselec MDS-2. As with other de-essing, this function can be tried in mid-side mode to see if it works best that way.

Analog Processing with Tubes or Transformers

Analog processing, especially "character" processors, can help with treating harshness.

Warming with a Compressor

High-quality compression can have the effect of adding warmth. This effect is increased when a side-chain mode is used. The bass is removed from the side-chained signal but is otherwise the same as the signal being processed. In this way, the compressor compresses when loud high-frequency sounds occur, smoothing them out. Analog compressors, such as the Pendulum OCL-2, are specifically known for what they do to smooth out high frequencies, especially when a side chain is used in this way. Multiband parallel compression also may be effective for this purpose.

Problem: Muddiness

Muddiness is quite a general description and could be caused by a number of problems, including masking to a level that cannot be fixed in mastering. Muddiness usually lives in the lower midrange and is addressed with tasteful equalization. The solutions run the gamut of equalization possibilities. Of course, the processing depends entirely on the recording at hand. As with many things in mastering, it is not always the problem range that is treated; sometimes working with other ranges can affect the problem range.

Problem: Part of the Frequency Spectrum Is Out of Balance During Loud Passages

Dynamic equalization allows some amazing feats of surgery to remove unwanted frequencies. Unlike multiband compressors, dynamic equalizers usually allow for the use of many different kinds of equalization curves. Examples include the Weiss EQ1-DYN and the Brainworx bx_dyn EQ. Dynamic equalization is preferred by mastering engineers over multiband compression for many tasks where either might seem applicable.

Problem: One Part of the Frequency Spectrum Is Too Dynamic

Multiband compression may be a solution when one part of the spectrum is overly dynamic or too loud for a short time. This might include an overly dynamic bass drum or vocal. These often can be tamed with multiband compression or dynamic equalization.

Problem: A Broad Part of the Frequency Spectrum Is Too High

When a broad part of the frequency spectrum is too high, a shelf or equalizer bell with a wide Q may be used. It may not always be used on the specific area. For example, if the lower-midrange and bass levels are too low, a high shelf cut might be used.

Problem: Less Than Full Sounding

When a mix needs "fattening," introducing distortion can provide an improvement. It is preferred to do this using analog equipment with "character," although there are some digital processors that perform the task well.

Some compressors can add a sound of fullness to a mix depending on the compressor being used and how it is used. The Manley Vari-Mu is certainly a unit known for its abilities with this type of processing.

Techniques for applying compression are covered in several places throughout this book.

Also, harmonic enhancement process such as the Sonnox Inflator, can help a mix to sound more full.

Problem: Fast/Medium Transient Sounds Stick Out Too Much

When fast/medium transient sounds are a problem, it is possible to digitally zoom in and fix them individually with equalization or volume automation. This can be the case with loud "s" sounds or other sounds that deviate markedly from the overall spectrum. Dynamic equalization is also effective here because it is possible to reduce a frequency only when it goes over a certain threshold.

Technique: Manually Reducing Peaks

If a large, single peak exists that is not essential to the emotion of a recording, manually reducing it can produce a better result than compression or limiting. This is typically accomplished using the volume automation that is a feature of virtually every DAW.

Technique: Reducing Level Before a Part Change

To give more impact to a part of a song, the level of the part before it can be lowered so that the change in level creates more impact. This is usually accomplished with volume automation, and the change in volume is usually done over time and in the most transparent way possible.

Technique: Analog-and-Digital Gain Staging

Gain staging is about setting gain controls within a signal path in a way that produces the best sound. Something as simple as gain staging can make a significant difference in sound quality.

Gain staging between digital processors is different from gain staging with analog processors. With digital processors, the main consideration is the differences between 16-, 24-, and 32-bit format.

16-, 24-, and 32-Bit Formats

At this time in mastering, the final output is almost always in 16-bit format. However, mastering engineers would never choose to process at 16 bits if it can be avoided. The only effective difference between bit rates is the signal-to-noise ratio. While noise is more carefully considered by most engineers at 16 bit, it is questionable as to whether it is ever audible, let alone a problem. However, at 24- and 32-bit rates, the signal-to-noise ratio is certainly a nonissue.

The 24-bit range is 256 times greater than the 16-bit range. It is an exponential increase in range. However, there are diminishing returns with respect to the perceptible difference in noise.

Essentially, 32-bit recordings are the same as 24-bit recordings; in fact, the range at 0 dBFS is exactly the same. The only difference is that 32-bit recordings have an extra 8 bits above 0 dBFS to prevent clipping from overs. This makes it virtually impossible for any digital clipping to occur at 32-bit resolution.

Digital Gain Staging

When working at 24 bits, the 16-bit level is located around –48 dBFS. Therefore, digital gain staging at 24 bits is not much of a concern. As long as the levels are decently above the –48-dBFS level, it should be fine.

When working at 32 bits, you must be careful because some digital processors only work at 24 bits. If there are overs (above 0 dBFS), there will be clipping if a digital processor in the chain is functioning at 24 bits. Because of this, when working at 32 bits, you must be sure that every digital processor used is capable of working at 32 bits (or be sure that there are no overs); otherwise, you must dither and convert to 24 bits.

Processing at 16 bits is not recommended during mastering. If provided with 16-bit source material with no other option, many engineers maintain high levels throughout the digital and analog signal chains. Upconverting to 24 bits does not change the range of the 16-bit recording and has no effect.

Analog Gain Staging

Various analog processors operate differently depending on the input level. Therefore, it is typically desirable to have gain controls before devices that work best at certain levels. Some units have gain controls built in, although some built-in gain controls may or may not produce gain that is pleasing. The Manley Vari-Mu is particularly known for its useful, character-adding gain sound.

A/D and D/A converters normally have a noise floor of their own; so many engineers will raise the levels digitally before going out to the converter. Before the D/A conversion, the analog levels are usually set to use the full range. For a clean signal path, it is a good practice to pass a nice and full signal both in and out of the converter.

Technique: Mastering with a Focus on the Vocal

The vocal is the central element, and mastering is best performed with this in mind. The vocal quality almost never should be sacrificed.

Technique: Working with a Vocal-Up Mix

A *vocal-up mix* is a version of a mix with the vocal boosted, perhaps 1 dBFS or more. Sometimes in mastering the level of the vocal will be affected. Some experienced mixing engineers will provide the mastering engineer with a vocal-up mix. In this way, if the processing begins to necessitate a vocal boost, the vocal-up mix can be used instead of using an equalizer to boost the vocals. Such equalizer boosts nearly always have side effects. Other alternative mixes include instrumental, acapella (vocal only), and TV mix (instrumental including background vocals).

Problem: Lacking Depth, Needs a Three-Dimensional Sound

Depth/three-dimensional (3D) sound largely comes from the skill of the mixing engineer. This is something that is to be preserved and embellished in mastering more so than added. There are a few methods, but they are lightweight in comparison with the possibilities in mixing:

- Expansion instead of compression can help to restore a sense of depth or three-dimensionality to a recording that is overly compressed.
- Mid-side equalization or adjusting the ratio of middle to side can help (e.g., raising the level of the side channel); often this is done with very small portions, perhaps only 0.5 dBFS.
- Skillful high-quality equalization can add to the depth of the sound. This is perhaps the most significant method on this list.
- Some compressors provide a subtle 3D effect, such as the Manley Vari-Mu.
- Parallel compression, in a few situations, can increase depth by raising ambient sounds.
- Use of a specialized stereo processor, for example, Bob Katz's K-Stereo or similar processors, may produce an improvement.
- Rare cases (usually not with a professional mix): Haas effect delays.
- Rare cases (usually not with a professional mix): Reverb (as described in a previous section).
- Rare cases (usually not with a professional mix): Left and right channels can be intentionally equalized differently to add a sense of depth or wideness to a mix. Sometimes the left and right channels are adjusted so that a cut in one channel is offset by a boost in the other. Different equalizers may be used for each channel that will impart a different character. This technique is seen almost entirely as a mixing technique, but there may be some mastering situations where it could produce a benefit.

Problem: Recordings Sound Different in the Car

In my experience, recordings only started translating to the car and other stereos in a fully satisfying way after upgrading to high-quality, full-range monitors (Tyler Acoustics D1s in my case). Also, using a high-pass filter set to remove most bass frequencies helps to audition how things may sound in a car—even when the car is full range. Auditioning with this high-pass filter in the studio can especially help to highlight loudness and tonal differences that arise in the car—it's almost like zooming in.

It has been said that the sound heard in the car can be emulated in the studio by listening at the opening of the room where the stereo channels have a somewhat mono presentation or by switching your monitoring system to play back in mono. The car listening environment does provide some stereo field perception, but it more or less tends toward a monophonic sound.

Even with the preceding techniques, it is important to listen in various environments, including a car. This is especially important when first getting started in mastering and when getting familiar with a new monitoring-system component or unfamiliar room.

Technique: Mixing Down to Tape/Mastering with Tape

Many engineers feel that there is no substitute for mastering from tape. It is common to see tape machines from manufacturers such as Ampex or Studer in mastering studios. Studer A80, A810, and Ampex ATR-102s are among the most popular tape machines used for mastering playback. Today, ¼-inch tape is the most popular, with ½-inch tape in second place. One-quarter-inch tape at a speed of 15 in/s (ips) is perhaps the most widely used today.

When mastering from tape, analog processing typically is performed between the tape machine and the A/D converter so that only one conversion to digital occurs. If a mixing engineer works in analog or uses analog processing on the mixbus, then recording to tape can prevent another conversion to digital before mastering takes place.

Layback is another possibility with tape machines. This is a mastering process where digital mixes are recorded to analog tape and rerecorded into digital. This is done to impart the qualities of analog tape into digital mixes. The various types of tapes have their own characteristics and a unique sound. Also, the saturation characteristics vary depending on the input level of the tape machine. The tape-machine components, calibration, and the machine itself play a role in the sonic characteristic it imparts.

When magnetic tape deteriorates with age, it is often called *shedding* or *sticky shed*. When this is a problem, tape baking is often used to prepare the tape for playback. When a tape squeals as it passes through the player or the playback sounds muffled or distorted, this indicates that baking is needed. Tape baking is not without risks; it can permanently damage the tape, so it must be done carefully. Also, if older tape is not baked, the friction from squealing is also not without risk because the tape oxide can tear off and permanently damage the tape. There are different methods for tape baking. It is typical to bake between 120 and 140°F for four to eight hours, commonly using a convection oven or food dehydrator. Gas ovens are never to be used for this purpose. An accurate thermometer or thermostat is a must. Acetate tapes and plastic reels are not baked because baking can damage both. Tape baking is only a temporary solution, although it can be done several times. The purpose of baking is to remove moisture, so, after baking, tapes often are sealed in a plastic bag with silica-gel packets to keep moisture down.

CHAPTER 8

Shaping Dynamics

The various mastering compressors have such different qualities. Some are loved for their punch, warmth, vintage tone, smoothing, general character, or maybe the way they treat high frequencies. Their strength also might be their functionality, to tame dynamics, and to minimize the effects of limiting. Every compressor is different and usually takes time to "get." Of course, it would take much more than a chapter to fully describe the nuances of every mastering compressor, so the common ideas are described here.

Compression is actually far less of a factor in today's professional mastering than most beginners would believe. There is so much hype surrounding compressors that it misleads many people about the significance of their role.

Of course, compression as we normally think of it is not the only dynamics processing used in mastering. There are several, and each type is discussed in this chapter.

Types of Dynamics Processing

There are several types of dynamics processing used in mastering, including:

- Compression (also called *downward compression*)
- Expansion (also called *upward expansion*)
- Parallel compression (also called *New York compression* or *upward compression*)
- Side-chain compression (alternate signal triggers compression action)
- Multiband compression (compression of specific frequency ranges)
- Limiting (a type of compression with a very high ratio)

Compression/Downward Compression

Downward compression is the most common dynamic range processing in mastering. Some mastering engineers rarely, if ever, use any other type, except limiting. With downward compression, the loudness of a recording is lowered when it exceeds a set threshold. Separate from compressing the dynamic range, there are other side effects called *character*.

Expansion/Upward Expansion

Expansion is when the loudness of a recording is raised when it exceeds a set threshold. It is rarely used in mastering, although some engineers may choose to use it with a recording that is overly compressed during mixing.

Parallel Compression

Parallel compression is performed by mixing a compressed signal in with its original. Parallel compression is used to effectively raise the lowest sounds. This process is usually best if applied subtly. It has been gaining in popularity in mastering and mixing. Some professional mastering engineers use it, whereas many prefer to use only downward compression.

Parallel compression can be accomplished manually by making a copy of a recording, ensuring that the original and copy are absolutely perfectly aligned. The copy then can be compressed and mixed in with the original. This should be done very subtly and can easily be as low as –20 dBFS or more below the original track, depending on the recording at hand. Parallel compression can be accomplished automatically by using the "Mix" knob on many plug-ins and some hardware units. The "Mix" knob is usually measured in percentages, with 5 to 20 percent often being the full range of possibilities for this technique in mastering. Because parallel compression is mixed in so subtly, it can involve extremely low thresholds. Of course, it should be set according to what pleases the ear.

Parallel compression is especially popular with acoustic or orchestral music and when reverb or low-level signals need to be raised.

Side-Chain Compression

Side-chain compression is when a signal is used for triggering the compressor action other than the signal being compressed. In mixing, side-chain compression is often used to compress one track according to the dynamics of another track.

In mastering, side-chaining is used a different way. The compressor will be triggered by an equalized version of the track being compressed, usually with bass frequencies de-emphasized. This is to avoid the compressor overly responding to the bass frequencies, which have the most energy and otherwise would dictate the compression action. Side-chaining is a somewhat popular option in mastering, although some professional engineers never use it or feel that they need it.

Multiband Compression

Multiband compression divides a signal into several frequency bands and applies compression to them separately.

Limiting

Limiting is essentially compression with an extremely high ratio—it is used to set a ceiling (a maximum level) that nothing should exceed. For the most in dynamics and depth, no limiting is used. When achieving loudness is a goal, limiting is virtually always in use.

Limiters impart their own character on the sound. Because of this, it is important to discover a limiter that works well with the other devices in the chain. Today, digital limiting, with look-ahead accuracy and high quality, is the standard. Digital limiters such as the Fabfilter Pro-L, Voxengo Elephant, and Slate Digital FG-X are popular. In the modern mastering studio, analog limiting has all but disappeared.

Compression Settings and Meters

Virtually every mastering compressor has four primary settings. These settings include the following:

- *Threshold.* A *threshold* is the level above which a compressor begins its action.
- *Ratio.* A *ratio* defines how much action will occur when the level goes over the threshold.
- *Attack.* The *attack* is how long the compressor takes to achieve its action.
- *Release.* A *release* is how long the compressor takes to end its action once the level has gone below the threshold.

Setting Attack/Release/Threshold

The settings of a compressor depend on the compressor itself and the recording at hand. As mentioned earlier, it takes time to "get" a compressor.

One general approach is to find settings that produce the most pumping, then set the attack and release times to work with the music, and finally, roll the threshold and ratio down to set the effect in with subtlety. A release time of 250 to

300 ms is often a great place to start. However, many engineers work with compressors much more intuitively than this approach.

Transients are an abrupt or sudden change in level, for example, we might think of a drum sound such as a snare as being very transient. Compressors affect transients when they are set to a fast-enough attack time. Slow attack times minimize the effect on transients. Negative effects on transients are usually easiest to hear when listening to the drums. If a compressor is having a negative impact on drum sounds, the attack time can be raised.

In mastering, a ratio of 1.5:1 is quite typical, while a ratio of 4:1 would often be considered high. An attack time of 0–10 ms is often considered fast, while an attack time of 30 ms might considered medium or slow. A release time of 100 ms would be considered very fast, while a release of 250–300 ms is quite typical. Compression settings depend on the material and task at hand, so more exacting compression recipes are avoided in this book.

Compressors are best set with a goal in mind, and achieving the envisioned goal may or may not involve any specific technique.

Macrodynamics/Microdynamics

The term *macrodynamics* refers to loudness over a long passage, primarily entire parts of a song, such as an intro, chorus, or verse. *Microdynamics* are faster sounds than macrodynamics. You might think of a drum sound or other fast sound. Microdynamics are mainly the transients and peaks, which have a much shorter duration. Both macrodynamics and microdynamics are considered in mastering.

RMS-Sensing Compressors

Root-mean-square (RMS) and peak sensing are two different methods compressors use to detect levels. Compressors that react to the average level are called *RMS-sensing compressors* and also may be called *averaging compressors*. The RMS/average level is more akin to how we actually perceive loudness when listening than peak level. The sound of RMS-sensing compression is usually more controlled. Because of this action, these compressors work best for macrodynamic compression.

Peak-Sensing Compressors

Peak-sensing compressors react to peak levels. Limiters, which are essentially compressors set with a very high ratio, are usually peak-sensing because they must quickly track the highest levels of a recording. Peak-sensing compressors are much more noticeable and less graceful than RMS-sensing compressors.

Compressor Response

Two different brands of compressors will produce different results, no matter whether they are both peak- or RMS-sensing—even with the same settings.

Character versus Transparent Compression

A processor that has *character* is one that imparts some tonality, as opposed to a *transparent* processor, which performs its function with little or no influence on the sound. Character compressors sometimes add their tonality even if no dynamics processing takes place. Other times character is imparted by various levels of processing. One must learn a compressor's nuances to skillfully apply its character. This is something that comes in time after experience with a compressor. Also, it is often discussed in forums, may be discussed in the user manual, or can be explored with a cordial call or e-mail to the manufacturer.

Transparent compression provides dynamics processing with very little harmonic distortion. Some units are capable of both character and transparent compression.

Mixbus Compressors

Mixbus compressors are used by mixing engineers on their master bus. Often they mix with the compressor on the bus during the entire mixing session. Mixbus compressors are usually ones that impart a considerable amount of character, whereas mastering compressors are generally more transparent. Sometimes compressors that are typically thought of as mixbus compressors are used in mastering for recordings that need more character.

Punchy Compression

This type of compressor can have a huge impact on the sound, and there are many factors that can cause this impact. For instance, the type of transformer can have a tremendous influence on tonality. Circuit architecture can make it punchier or less punchy. For example, an API 2500 is great for fast or aggressive music and adding punch while also remaining quite versatile. In contrast, the Pendulum ES8 provides a very hi-fi sound but is less punchy and is not as fast.

Serial Compression

Serial compression is the use of two compressors. Sometimes the purpose will be to combine the characters of several units. Also, different compressors have different strengths, and sometimes combining two can be just the right fit. For example, one compressor might be set for microdynamics processing, whereas another is set for macrodynamic processing. It is very common to use one compressor to tame peaks, with a fast RMS- or a peak-sensing action, along with another that is used with slower settings and with an RMS/averaging sensing action.

Volume Automation for Macrodynamic and Microdynamic Adjustments

Sometimes, instead of using a compressor, volume automation can be used. In many situations, this can provide a more transparent adjustment than using a compressor. Whether working to fix a specific microdynamic or macrodynamic problem, this can be the best option. Remember, the action of a compressor is intended to be an automated loudness adjustment—it also can be done manually. For example, if a chorus is too loud in comparison with a voice, instead of working with an RMS-sensing compressor with slow attack and release times to even out the difference, a simple volume-automation curve could be used. Also, if there was a kick drum within a song that accidentally rose high above the rest of the recording, if a remix is not an option, a simple volume-automation curve might produce better results than a compressor.

Gain-Reduction Meters

Most compressors have a meter that shows the gain reduction taking place. In mastering, often compression is used subtly. Gain-reduction meters may not show any or only very slight movement, even when the compressor is optimally set.

Compressor Input Levels

Some compressors, such as the Manley Vari-Mu, have an input control that can have a big impact on the sound. With the Vari-Mu, if the input level is raised compared with the output, the tubes are driven harder, and more character is added.

Compressor Output/Makeup Gain

The output/makeup gain can be used on a compressor so that the input and output signals are at the same level. In this way, the effect of the compressor can be auditioned using the bypass feature. It also avoids the difference in loudness affecting the engineer's perception of the processing. Makeup gain also can be used to simply add gain into the signal. This must be done carefully because the makeup gain of different compressors can sound much different.

Linked versus Unlinked Compression

Linked and unlinked modes on a stereo compressor indicate whether the compressor will have the same or different action on each channel. Linked compression mode allows the compression to be more apparent, which may be useful when a compressor has a certain character. You should always try linked and unlinked modes to discover which sounds better.

Expansion/Expansion Before Compression

A few engineers will use expansion processing to expand the dynamic range or use a transient designer process to highlight the transients and increase the dynamic range of a recording before using a transparent limiter to compress it again. This technique is not very popular or well accepted, although it is intended to add punch and expand the sound stage.

Analog Compressors

There are several different types of analog circuit methods used for compression. The types include voltage-controlled amplifier (VCA), opto/ELOP, variable mu, field-effect transistor (FET), and pulse-width modulation (PWM). Each of these gain-control mechanisms has its own characteristics.

Voltage-Controlled Amplifier (VCA)

The term *VCA compressor* is used when integrated circuits (ICs) are used to control the gain reduction. Technically speaking, virtually all analog compressors could be called VCA compressors because voltage is being controlled, but in studio parlance, it refers only to ones that use ICs. VCA compressors are known for being very fast and are used on the fastest-tempo rock and metal, along with many other styles, and can work with slow tempos as well.

Opto/ELOP

Opto compressors used for tracking and mixing are quite slow. Opto compressors created for mastering are much faster, including the Pendulum OCL-2 and the Shadow Hills Mastering Compressors' opto side. Opto compression is normally talked about as having a smoothing effect, especially smoothing the high frequencies, with lots of "mix glue" while adding a subtle amount of fullness. Their action is based on a light-controlled optical element to produce smooth and pleasing gain reduction.

Variable Mu

Variable mu compressors are thought of as having more character, as well as providing smoothness, depth, and fullness. With this type of compressor, gain reduction is accomplished with a vacuum tube.

Field-Effect Transistor (FET)

FET compressors are not as widely used in mastering. Often FET compressors are thought to impart too much tonality for mastering.

Pulse-Width Modulation (PWM)

PWM compression is typically used to emulate the types of compression listed earlier.

CHAPTER 9

Achieving Loudness

This chapter is about achieving loudness—ideal loudness, extreme loudness, and everywhere between. The subject is actually quite sensitive in mastering because of extreme loudness trends called the *loudness wars*. As a result of the loudness wars, the loudness of recordings has been pushed in a way that has seriously degraded recording quality. When mastering is performed best, recordings are set to a loudness level that is the highest possible without negatively affecting tone or dynamics. However, mastering engineers may be called on to produce loudness anywhere from ideal to extreme, so it is an important part of the mastering knowledge set.

Apparent/Perceived Loudness

The way the human ear perceives loudness is quite complex. When you view a waveform in a digital audio workstation (DAW), you are seeing the peaks and troughs of the pulse-code modulation (PCM) wave—a very poor visual representation of perceived loudness. No meter fully replicates it, although root-mean-square (RMS)/average meters set to slow response can do a somewhat decent job but are still lacking. LUFS meters are the most accurate meters for measuring loudness. The most useful LUFS meters are typically those that include a display of recent levels. In mastering, the short term (S) and loudness range (LRA) modes are the most common. However, most mastering engineers set the final levels by ear—which is always the most accurate tool.

Beyond Ideal Loudness, There Is Quality Loss

Pushing the loudness beyond its ideal level causes a loss in sound quality. Pushing it to the extreme can cause severe distortion. When higher-than-ideal loudness is demanded of a mastering engineer, everything must be done to preserve the original tonality of the recording as much as possible. It often helps to compare the original and the processed versions at the same level to get a sense of the impact of the processing.

When Did the Loudness War Begin?

Many people believe that the loudness war was something that has come along within the past two decades with the proliferation of digital recording. The fact is that the loudness war has been going on since the days of vinyl. Some engineers in pursuit of loudness on vinyl would push the levels so high that listener's needles would physically jump out of the groove!

Why Does the Loudness War Exist?

The loudness war is mostly due to the fact that people associate higher loudness with a more intense listening experience and higher fidelity. Producers want to leverage this fact to compete, so they want their recordings to be louder than others to gain an edge.

With radio processors such as the Orban and Omnia processors, recordings mastered to be louder will not be louder compared with any other recordings when played on the radio. However, if someone is listening to a recording on an MP3 player, smart phone, or playlist, the variation in loudness will be heard and will likely affect the listener's perception.

Future of the Loudness War

Judging from the past, we might assume that the loudness war will go on forever, as it almost always has. There is one movement that challenges this future. Apple has begun to integrate a feature called Sound Check into iTunes, iPods, and iPhones that detects the average loudness and adjusts the volume so that no matter how much limiting is applied, all recordings will play back at generally the same loudness. Mastering engineer Bob Katz has worked diligently for many years to get manufacturers to implement these types of processes. They would spell the end of the loudness war and ultimately would lead to better sounding recordings.

Loudness Potential of a Recording

Some recordings have a higher potential for loudness than others. High-quality recordings, when degraded with loudness techniques, sound better than recordings that are not as high quality to begin with. The loudness potential of a recording is something that starts with production and mixing.

Digital Clipping

Digital clipping occurs when a signal attempts to go over the highest possible level—0 dBFS. When clipping occurs, there will be two or more samples that will be at the same level. Severe clipping usually involves four or more samples at the same level. The sound of severe digital clipping is very harsh. While working to achieve loudness, perhaps the most basic goal is to avoid harsh digital clipping artifacts.

Using Clipping

Clippers, such as gClip, as well as clipping features in some limiters, will allow a type of wave shaping that can sound more natural than some types of limiting depending on the material being clipped. Usually this is used when there are only few very fast peaks, where a shaped clip might affect the sound less than the action of a limiter. This type of process has its own character.

Compression Before Limiting

When trying to achieve loudness, it sometimes can help to compress before the limiter. There is a balance that should be achieved. Often, if the master must be very loud, little compression might be used, and the limiter will be used for the majority of the increase in loudness. Using too much compression can take away from the tonality that will already be reduced by the limiter. Sometimes it can be best to use no compression at all.

Digital Limiting

Digital limiters are one of the most important tools for maximizing loudness, and each one has its own character. Limiters such as Fabfilter's Pro-L have different character options that allow for a diverse selection of sounds. Digital limiters such as the Ozone Maximizer's IRC III and the Voxengo Elephant have amazing transparent qualities. Digital limiters have a major advantage over analog limiters because they can have look-ahead features that allow them to quickly react to a transient, even before it has occurred.

Operating a Limiter

Limiters are simple to operate. There is almost always a gain control that is raised until the desired level of loudness is reached. Limiters also usually have a ceiling control that allows the user to set the highest-allowable peak. Limiters with a slow attack may exceed the set ceiling because the limiting action will not react quickly enough to grab fast peaks. Also, today's limiters have dithering and oversampling built in so that separate processors will not have to be used for these features.

Sensitivity of the Ear

When maximizing loudness, it is important to consider the equal loudness contours (or their predecessor, the Fletcher-Munson curves). These indicate that the ear is most sensitive around the 3-kHz region. Therefore, boosts in this area or cuts in areas below it will raise the perceived loudness. This frequency range requires less energy than the bass or low-middle frequency ranges. As such, the loudness can be raised even higher with a limiter, without clipping, when the 3-kHz range is boosted. However, this kind of approach must be taken in a way that minimizes ear fatigue for the listener. Boosting loudness with this approach is used when the most extreme level of loudness is demanded. A very slight adjustment around 3-kHz goes a long, long way.

Clipping a High-Quality A/D Converter

A common technique to achieve loudness is to raise the gain of a signal in the analog domain before digital-to-analog (D/A) conversion. Essentially, high-quality A/D converters act in the role of a limiter when used this way. Of course, this requires a high-quality gain control to be somewhere within the signal path. Some engineers may use this only for a very light clipping effect while relying on a digital limiter afterward for the heavy lifting. In this way, if the client requests a change in loudness, it can be accomplished without the time required for analog processing. Also, some engineers never use converter clipping, especially given the quality of modern digital limiters.

Serial Limiting

This is a technique for achieving loudness in which several limiters are used at different stages, each taking a little bit off the top to achieve loudness. The idea is that several processors, with their various compression/limiting actions, can achieve loudness better than a single processor. Of course, there can be diminishing returns with the number of processors used for this approach. One prime example is to use a small amount of transparent digital limiting or compression before converting to analog, slightly compress in the analog chain and gently clip a high-quality A/D converter; and finally, use a high-quality digital limiter once back in the digital domain. At each stage of this approach, just a little more limiting/compression is applied for the purpose of achieving the most transparent loudness.

Digital Limiter Ceiling of –0.3 dBFS

Using a ceiling of –0.3 dBFS for limiting is virtually a mastering standard. This ceiling helps to prevent consumer D/A converter clipping and intersample peaks. This ceiling can be set with a limiter or maximizer. Instead, it is possible to set a digital audio workstation (DAW) master fader, track fader, object fader, or a transparent gain control after the limiter to –0.3 dB while leaving the limiter set to 0 dBFS. While this is very widely accepted, a few mastering engineers may use a ceiling of –0.1 dBFS, although there is virtually no audible difference.

NOTE *In this section I am discussing –0.3 dBFS, not –3 dBFS. There is a big difference!*

Maximizers/Multiband Limiters/Inflators

A *maximizer* is usually a limiter that involves multiband or other "smart" processing. Some processors designated as limiters could be called *maximizers*, so the terms aren't definitive. For example, the Slate Digital FG-X responds to transients differently depending on frequency, but it is still called a limiter. The Waves L3 is a well-known multiband maximizer. The iZotope Ozone maximizer provides great results and is extremely transparent. Normally, a maximizer is used for the same function as a digital limiter.

Loudness can be raised in ways other than with dynamics processing with processors such as the Sonnox Inflator, Voxengo VariSaturator, and Universal Audio Precision Maximizer. The Sonnox Inflator has gained the highest popularity for this role and some see it as a "secret weapon" for loudness and thickening. When it is used, often the "curve" is adjusted with the effect at 100 percent, then the percentage is lowered to mix in the effect. However, there are many mastering engineers who do not use this type of processing. When it is used, it is almost always just before the limiter.

Ideal Loudness

Ideal loudness is set by ear and according to taste. Some engineers use a target RMS level, such as –10, –12, –14, or –20 dbFS RMS. One of the most noted proponents of ideal loudness is Bob Katz, who declared a system for guiding toward ideal levels. Essentially, it is a system where recordings are set to a standard RMS level of –12 dBFS RMS for broadcast/radio, –14 dBFS RMS for pop/rock/country, and –20 dBFS RMS for film/classical/hi-fidelity, with the peaks of each at –0.3 dBFS. His K-Meter shows RMS levels and peaks to make this easy. The K-System is discussed in more detail later in Chapter 11.

Processing While Focusing on the Loudest Passages

Working with a focus on the loudest passages in mastering can help to quickly find the optimal loudness potential of a recording.

True Peak Ceilings

When working with recordings intended for broadcast, especially television broadcast, and the broadcaster requires adherence to the Advanced Television Systems Committee (ATSC) A/85 RP, European Broadcast Union (EBU) R128, and International Telecommunication Union ITU-R.BS.1770 standards, then the true peak level must be visualized and set. There are many true peak meters available, including the Waves WLM. Limiting in a way that conforms to the true peak specifications requires limiter oversampling of 4x, and it is usually a good idea to use a limiter with intersample peak protection and a ceiling of at least –0.1 dBFS, although –0.3 dBFS is widely thought to be best.

Broadcast Loudness Standards

In some areas of broadcast, there are recommendations and legal standards on loudness. These include

- Advanced Television Systems Committee (ATSC) A/85 RP
- European Broadcast Union (EBU) R128

- International Telecommunication Union ITU-R.BS.1770
- Commercial Advertisement Loudness Mitigation (CALM) Act

In the United States, broadcast standards are mainly for television, whereas in Europe, the EBU R128 covers radio, television, and other electronic media. However, EBU R128 is not adopted into law, although a vast majority of European broadcasters adhere to it and require the recordings they broadcast to adhere to it.

ITU-R BS.1770

International Technical Union Recommendation Number BS.1770 outlines many standards for broadcast audio. It defines loudness unit (LU) and loudness unit referenced to full scale (LUFS), true peak, foreground loudness, and dialog loudness as incorporated into the EBU R128. The Dolby Media Emulator is software designed to help with this metering and other film/TV related monitoring tasks.

EBU R128

European Broadcasting Union Recommendation Number 128 (EBU R128) is a recommendation to European broadcasters regarding loudness. It largely incorporates the ITU-R BS.1770 standards for measuring loudness, including its definition of loudness unit (LU) and loudness unit relative to full scale (LUFS), true peak, foreground loudness, and dialog loudness. It recommends a standard of –23 LU/LUFS (with a deviation of –1.0 LU if exacting normalization is not practical, such as with live situations), with a maximum permitted true peak level of –1 dBTP. Also, it recommends matching the loudness of a commercial with the "foreground loudness" of its associated program. It recommends noting a recording's maximum LU/LUFS in its digital metadata. Whereas these are the main recommendations, there are others, mostly that define how the meters for making these adjustments should be fashioned. There is also "loudness meter test material" that has been released and updated by the EBU for ensuring that meters used for this purpose are calibrated accurately.

ATSC A/85

This standard is comprised of recommended practices developed by the Advanced Television Systems Committee (ATSC). It involves many of the same topics as ITU-R BS.1770 while providing a comprehensive background of information surrounding relevant topics.

CALM Act

The Commercial Advertisement Loudness Mitigation (CALM) Act was passed by Congress and signed by President Barack Obama into U.S. law on December 2,

2010. The CALM Act adopts parts of the ATSC A/85 standard into law. This law concerns audio used for television commercials. It limits the loudness of audio used during commercials, making it in line with the loudness of its associated program. It accomplishes this by linking the commercial's loudness with the program's using an *anchor*. The anchor is usually either the dialog of the program or what is called the *foreground loudness* of the program.

CHAPTER **10**

Fades, Sequencing, and Spacing

The way you get from one song to another on an album is important to the listening experience. You feel it. In mastering, you should always work to achieve the best effect as you transition between songs. This transition is made up of not only the fades and spacing but also the sequence of the songs.

Spacing Between Songs

The spacing between songs is important to the listening experience and must be set in a way that works artistically. The way you accomplish this varies depending on the digital audio workstation (DAW) you are using.

With some DAWs, spacing between songs is very easy to set and audition. Samplitude and Sequoia are two DAWs that make the task easy. Some of the least powerful mastering DAWs require the engineer to master a recording, bounce it to a wave, and then use another program to audition the spacing. Being able to quickly audition and adjust spacing is essential for creating a musical and natural flow.

It is usually best to audition the space between recordings by listening to playback from a point that is significantly before the space. This helps to gain perspective about how the listener may experience it. Perfect spacing is a matter of timing and musical intuition.

Most DAWs feature a mode where the play cursor stays at its current position when playback is stopped. Otherwise, play cursors normally return to the starting point when the user presses "Stop." This mode can be useful for creating natural-sounding spaces between tracks. Assigning a keyboard shortcut to this mode adds convenience.

There exists a technique where a *room-tone* recording may be used in the spaces between recordings, where otherwise the listener would hear perfect digital silence. This creates continuity between songs. Room tone is extracted from the silence at the head or tail of a recording or by requesting a clip from the mixing engineer if no head or tail provides long enough silence.

Performing Fades/Cross-Fades

Fades and cross-fades are also accomplished in ways that work artistically with the recording at hand and are very DAW-specific. It is important to explore all the fading options of the DAW used for mastering.

Cross-fades are used to fade between one recording and another and are often used to make seamless transitions between songs. When this is done, the CD pause time is set to zero, and the track marker between the songs is set wherever desired.

Fades and cross-fades are usually done nondestructively, that is, in a way that does not alter the original source.

Fades During Mixing Stage

Fades during the mixing stage are performed in a DAW with automation, fade curves, or a master fader. Fading during the mixing stage provides a big benefit—it allows the creativity of the mixing session to be put into the fades. Of course, this must be done carefully. Today, performing fades during mixing is most common.

Fades During Mastering Stage

There are a few different ways to request fades during mastering. First, they can be discussed and performed during attended sessions. For unattended sessions, fade times and lengths can be provided to the mastering engineer. Sometimes both faded and unfaded recordings are sent for mastering, allowing the mastering engineer to work from a reference fade. Also, it can always be left up to the mastering engineer to perform fades at his or her own discretion.

Performing Fades During Mastering Has Its Own Set of Benefits

Most dedicated mastering DAWs have more fade types and fade customization than typical mixing DAWs. Mastering fades are done in sequence after the limiting and compression so that they are not affected by those processes, as are fades performed during mixing. Performing fades during mastering is the very best choice when the transitioning and sequencing of the album will be highly creative.

Sequencing an Album

Today, the sequence of songs on an album is usually decided before mastering. However, mastering engineers usually offer suggestions if the album sequence seems problematic. Rarely, the mastering engineer will be called on to set the sequence. When this is the case, the songs are usually set according to the mastering engineer's musical intuition.

Noise Reduction

Noise-reduction processing is considered to be *corrective* processing and is usually performed at the beginning of the chain.

Denoise First

Noise reduction and other corrective processing are usually done very early in the chain. Usually, they are preceded only by DC-offset removal, if needed.

Noise at Beginnings and Endings

Noise is usually most audible at the beginning or ending of a recording. It is usually best to process only the sections where it is audible. Often this processing will be performed destructively.

Learn Feature

Many noise-reduction processors are equipped with a *learn feature*. A playback loop is set on a section where only the noise can be heard. Then the processor is set to "learn." Afterwards, the noise-reduction processing is applied with the noise profile it learned. This processing varies depending on the software being used, so reading the user manual is a must.

Noise/Hiss/Clicks/Pops

Different noise-reduction algorithms specialize in various types of noise problems. The correct processing should be used for the problem at hand.

Spectral Editing

Spectral editing processors allow for noises, clicks, and pops to be removed using a spectral display. At its core, this process is an automated equalizer. Samplitude and Sequoia have spectral editing built in, allowing for surgical reduction of clicks and pops in an unnoticeable way.

Careful Application

Noise-reduction processing can easily degrade a recording, so it is very important to apply it carefully.

CHAPTER 11

Visualizations/ Metering

T here is a story about Mitch Miller, a figure in American music, who once became frustrated with an engineer paying too much attention to the meters, so he placed tape over all the meters in the studio. As you do your mastering work, you must rely on your ears, not your eyes. For novices, this is a real challenge because it takes time to develop a sense for mastering. This leads many who are just starting out to try to find systematic, often visual ways of working. For instance, they might try a curve-matching equalizer or maybe some metering system that promises better results. There won't be such a visual approach to mastering any sooner than there will be such an approach to other aspects of creating music, such as songwriting or production. With that said, there are a few benefits visualizations can provide, but they are mostly for details that do not involve tonality.

FFT/Fourier

Every visualization that shows frequency content is based on a mathematical operation called a *Fourier transform*. It all starts with a waveform/signal. Using it as an input, the Fourier transform is calculated, allowing for the frequency spectrum to be visualized. It would be impossible to make such a calculation instantaneously because it takes some time for frequencies to oscillate. This means that Fourier transform visualizations are only an approximation.

You see Fourier transform visualizations rendered in many different ways, some that look more impressive than others, but essentially they are based on the same concept. Fast Fourier transforms (FFTs) are Fourier transforms that have an optimized algorithm so that they can be calculated much faster than a Fourier

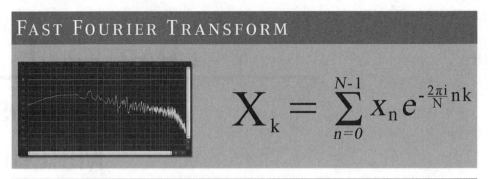

FAST FOURIER TRANSFORM

$$X_k = \sum_{n=0}^{N-1} x_n e^{-\frac{2\pi i}{N} nk}$$

FIGURE 11-1 The fast Fourier transform (FFT) is familiar to most people as a display of the frequency energy of a recording, as shown here.

transform normally would take to process. An example of the FFT and its formula is shown in Figure 11-1. As of January 2012, there is a new, even faster Fourier transform called the *sparse fast Fourier transform* (sFFT) that has yet to be widely implemented.

Using Spectral Analyzers

There are two types of spectral analyzers common in mastering: spectrograms and spectroscopes. Such analyzers can help to locate a frequency heard with the ear and can be an alternative to the technique of sweeping with an equalizer and listening. Typically, a spectroscope will show a downward-sloping trend for most masters. Spectrograms often show the frequency energy represented with colors. Some use the frequency spectrum of light (ROYGBIV) to display frequency intensity. Others use their own unique color palettes. Because bass frequencies use the most energy, audio spectrograms for most recordings have a gradient fade from most intense to least intense. Some novice engineers will recommend making adjustments in a way to make the visualizations display have a smooth slope or gradient. It must be reiterated that you must work according to your ears, not your eyes. Examples of the spectroscope and spectrogram are shown in Figure 11-2.

Bit Meters

These meters display the bit rate of a recording. Sometimes, when you think that you may be working at 24 or 32 bits, you actually may be working at 16 bits—accidents can happen. Bit meters keep you apprised of the current bit depth. It is similar to a car overheating light; you may not need it often, but it can be helpful when something goes wrong.

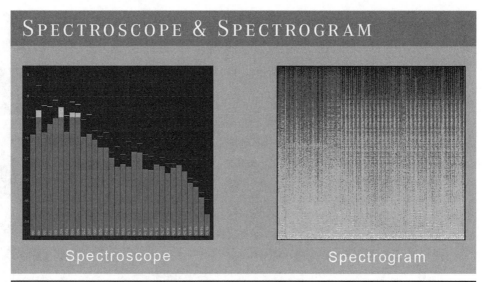

FIGURE 11-2 The spectroscope and spectrogram are common audio visualizations.

Correlation Meters

Correlation meters show the similarity of the left and right channels. When the left and right channels are the same (mono), then a correlation meter will show +1. When the left and right channels are exactly out of phase, the meter shows –1. Typically, most recordings are between 0 and +1 and most often are between +0.5 and +1. Most mastering engineers work without using correlation meters. In mastering, it is normally the ears that are relied on for making any adjustment to stereo width. Typically, listening in mono is a better approach than using a correlation meter for detecting phase problems, except in lacquer cutting where a phase correlation meter can be useful. An example of a correlation meter is shown in Figure 11-3.

Vectorscope

Vectorscopes are a type of oscilloscope used with stereo recordings. They are similar to correlation meters, although they show more about the stereo field. When they are completely vertical, it indicates that the source is mono. While a circular shape is sometimes said to be the ideal stereo shape, there is actually no ideal shape. An oval shape that is horizontal could possibly indicate a width that is too wide or a phasing issue. A horizontal shape indicates that the left and right channels are out of phase with each other. An example of a vectorscope is shown in Figure 11-3.

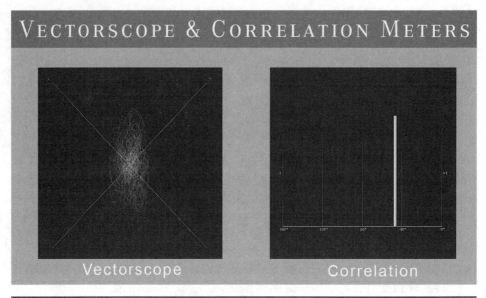

FIGURE 11-3   Vectorscope and correlation meters display information about the stereo field.

Reconstruction Meters

Reconstruction meters display the levels of a waveform after it is converted to analog. Their primary use is for the detection of intersample peaks. Most mastering engineers take steps (previously discussed in this book) to guard against intersample peaks, making this type of metering unneeded. Solid State Logic (SSL) offers a free reconstruction meter called X-ISM for detection of intersample peaks.

Meter Action/Speed

There are various kinds of meters that can be found in both hardware and digital audio workstations (DAWs), each with different operation. Almost everyone is familiar with VU meters (volume-unit meters), which are fairly slow in an attempt to mimic the way that humans perceive loudness. There are also peak program meters (PPMs), which show audio peaks and are much faster than the human ear. Because of this, PPMs provide very little insight into perceived loudness. VU meters only vaguely represent perceived loudness but are closer than PPMs. Today, LUFS meters are the closest representation we have of loudness using a meter. As always, the ear is the best tool for the job.

CHAPTER 12

Preparing the Final Output

The mastering role is nearly complete. It's time to put the recordings in their final format, ready for delivery to the client. These formats include CD, WAV, MP3, and a few others. You should take great care in accomplishing this final step because you are creating something that will be used for manufacturing or distribution. Ultimately, the result of this step will shape the client's perception of your mastering services.

Providing Client a Preview for Approval

Clients will need to listen to the final master for approval. MP3, WAV, DDP, or reference CDs can be used for this purpose.

Mastering studios will often send a WAV file to clients for approval. However, using WAV files does not allow previewing of album cross-fades, CD pauses, CD text, and International Standard Recording Codes (ISRCs). Instead, you might use DDP files. DDP files allow for everything to be previewed and are the most complete way to have a client approve the work. Software such as Sonoris DDP and Tonic DDP allow mastering studios to send clients a licensed DDP player for previewing. With this option, the client may not need to be shipped any physical disc, saving shipping and supply costs. If the client approves, he or she can simply send the DDP on to the duplicator/replicator, and the order is complete.

Another option is to provide the client with a reference disc. This is the slowest option if the client is not local.

Previewing is important because clients can listen in a familiar environment with familiar equipment. Afterwards, the client can provide feedback and make requests for any revisions.

Making Revisions

Changes are made at their appropriate processing stage. For example, if the client would like a brighter sound, equalization would not be applied to the final master; it should be adjusted in the chain in sequence. Once revisions are made, the client is provided with another preview.

Quality Control

The final master is listened to entirely, sometimes in headphones, for quality control. It is best to listen to the entire final output that will be duplicated/replicated, whether it is a DDP or a CD. The heads and tails of the recordings are where problems occur the most. Careful attention is paid to these sections. Quality-control responsibility can be shared with the client by asking him or her to carefully listen to the final output before duplication/replication.

Red Book/Rainbow Books

There are a number of standards such as the well-known *Red Book* standard. Each has its own color, and collectively they are called the *rainbow books*, as shown in Table 12-1. The rainbow books define the various formats for CDs.

TABLE 12-1   Rainbow Books

Red Book	CD-DA (digital audio, extended by CD-Text)
Yellow Book	CD-ROM/CD-ROM XA (read-only memory)
White Book	VCD (video), CD-Bridge (hybrid), SVCD (super video)
Blue Book	E-CD (enhanced CD), CD+ (CD plus), CD+G (plus graphics), CD+EG/CD+XG (extended graphics), also known as CD extra
Beige Book	PCD (photo)
Green Book	CD-i (interactive)
Purple Book	DDCD (double density)
Scarlet Book	SACD (super audio CD)

It is good for mastering engineers to be aware of these standards and remain familiar with the *Red Book* standard. Rarely, you may be called on to create an enhanced/mixed-mode CD, which is defined by the *Blue Book* standard and requires specialized software. For mastering engineers, the details of the *Red Book* standard mostly apply when dealing with CD duplicators/replicators. If their software returns an error, you may need to troubleshoot to discover the cause, requiring knowledge of the *Red Book* standard.

Red Book CD Specifications

There are a few basic specifications for a *Red Book* CD that every mastering engineer should know:

- Maximum running time is 79.8 minutes.
- Minimum duration for a track is 4 seconds, including any pauses.
- Maximum number of tracks is 99.
- Maximum number of index points (subdivisions of a track) is 99, with no maximum time limit.
- It is in 16-bit format with a sample rate of 44.1k.

Setting Track Markers

Track markers are set between each recording of an album. This is where the CD player will begin playing when a track number is selected. Track markers must be set on a CD frame (there are 75 per second); otherwise, the marker will be moved to the closest frame. To do this in Samplitude/Sequoia, the digital audio workstation (DAW) is set to "CD MSF" (minutes, seconds, frames) mode, and the snapping is set to frames before the track markers are set.

CD Pause Length

A pause length can be set for a CD so that between each track, the CD player will pause for that length. This is the familiar 2-second gap between songs on some CDs. In recent years, this pause length is often set at zero, especially when there are cross-faded songs that require no gap.

Today, virtually all engineers prefer to set the overall pause length to zero and perform spacing with other means. This gives the benefit of more natural transitions, and spacing is retained when a CD is ripped.

The pause before the first track is something that can be set independently of the pause between all other songs. The *Red Book* standard requires that this pause be at least 2 seconds. Most mastering engineers always use that minimum length.

Track Offsets

Track offsets are the space between the track marker and the beginning of the recording. Often some amount of silence is left between the track marker and the beginning of the music, or it is inserted with a delay. Many mastering engineers measure the track offset in CD frames. There are 75 CD frames per second. This unit of measure is commonly abbreviated as CDF and is designated in some software as CD MSF (this stands for compact disc minutes, seconds, and frames). Perhaps the most common track offset is 6 CDF (80 ms), with 15 CDF (200 ms) also being common.

This offset is easily accomplished with a DAW's built-in offline/destructive delay processor, which is capable of precisely inserting this delay. First, a cut would be made before the first oscillation, and then the delay would be processed. This feature is not offered with any plug-in because of the way many DAWs perform delay compensation. You must use the DAW's built-in delay processing. Of course, this spacing would not be used when a song is being cross-faded with another song.

International Standard Recording Codes

ISRCs are like social security numbers for recordings—each one is unique. However, they are not issued by a government but instead are an agreed-on international standard administered by the International Confederation of Societies of Authors and Composers (CISAC) in France. CISAC appoints creative societies to facilitate assigning codes in various countries, such as the Recording Industry Association of America (RIAA) in the United States. ISRCs can be obtained by the client and supplied to the mastering engineer. Clients can obtain the codes from their record labels, digital distributors such as CD Baby and Tunecore, or apply for their own ISRC prefix (in the United States, the prefix would be assigned by the RIAA). It is also possible for a mastering studio to become an ISRC manager, although the vast majority of studios do not because of the paperwork, cost, and responsibility of record keeping.

MCN/UPC/EAN Codes

When Universal Product Code (UPC)/European Article Number (EAN)/barcodes are embedded into an audio CD, they are called *Media Catalog Numbers* (MCNs). MCNs are 13 digits, and EANs are also 13 digits. UPCs are 12 digits, so a zero is added to the beginning of the 12-digit UPC to make it an MCN when encoding to CD-Text. Sometimes these codes may be called a *matrix number* or *release number*.

These codes are unique for a product being sold, such as an album, extended play (EP) or even a single. Ideally, these will be supplied by a record label. Independent artists usually obtain them from digital distributors (often CD Baby or Tunecore) or their CD manufacturer. Clients also can apply to obtain UPCs from the Global Standards One (GS1), although the price is a minimum of $760. GS1 is no longer offering the ability to resell UPC codes and has recently implemented initiatives to stop businesses from reselling them. It is generally allowing people who were already reselling the codes to continue, considering them to be "grandfathered," including CD Baby, Tunecore, and others.

CD-Text

CD-Text is digital text embedded into a CD that is displayed by some CD players or used to provide information to databases regarding a CD. The CD-Text format was jointly created by Phillips and Sony. Both companies hold patents for CD-Text, and if CD-Text is used, royalties must be paid to Phillips under their "Joint CD Disc Patent License Agreement" (search Google for more information). CD manufacturers may be billed a licensing fee for creating discs that use CD-Text, and normally, this fee is incorporated into their pricing or passed on to labels. When an album is projected to have high sales, then consideration should be given as to whether CD-Text should be used. The licensing fee is $0.03 per disc. For example, an album with 500,000 copies pressed would have a CD-Text licensing fee of $15,000. For a run of 1,000 copies, it would be only $30.

CD-Text is being supported by fewer and fewer CD players. Also, an increasing number of media players draw their information from online databases instead of CD-Text. Because of this and the licensing fees, CD-Text is becoming less common, with many artists and labels avoiding it altogether.

CD-Text should be verified as correct before creating the final output for the client. It is usually best to copy and paste it from client-submitted information. Printing a PQ sheet and having the client look it over is a great way to avoid errors.

The CD Text fields are

- Arranger [Name(s) of the arranger(s)]
- Composer [name(s) of the composer(s)]
- Disc ID (disc identification information)
- Genre (genre identification and genre information)
- ISRC (ISRC of each track)
- Message (message from the content provider and/or artist)
- Performer [name(s) of the performer(s)]
- Songwriter [name(s) of the songwriter(s)]

- Title (title of album name or track titles)
- TOC Info (table of contents information)
- TOC Info2 (second table of contents information)
- UPC/EAN/MCN (UPC/EAN code of the album)
- Size Info (size information of the block)

Premaster CD

If a mastering studio creates a CD for duplication/replication, it is technically called a *premaster CD*. It is a CD from which a glass master can be made for stamping CD copies in replication. A company named Sonic Solutions created a format called *premaster CD* (PMCD) at one time that was later discontinued. At that time, there was a distinction between PMCDs and *Red Book*–compliant CD-Rs intended for duplication. Now any *Red Book*–compliant audio CD-R intended for duplication is generally referred to as a *premaster CD* (PMCD).

Writing Speed

Premaster CDs are normally burned at a low speed, with no other processes running. The slow speed and minimal central processing unit (CPU) activity can help to reduce write errors. A speed of 4x is common but doesn't necessarily result in the lowest error rate with every burner or CD-R medium. It is important to test to find if any of these factors affect your system and to test different speeds to find which gives the lowest errors.

Disk-at-Once/Track-at-Once

In many professional mastering applications, selecting the write method is not an option. For those where it is an option, the *Red Book* standard requires disk-at-once (DAO). The alternative, track-at-once, is mostly for archival files and is created with software not typically used for mastering.

Error Checking with Plextor/Plextools

Hardware error checking with Plextools is by far the most popular method for performing CD error checking. There is a software error checker included with Nero Burning ROM, but it is not as accurate. Plextools is now free, but it must be used with a Plextor CD burner that is compatible with the error-checking features of Plextools. These Plextor drives include

- Plextor 716A (IDE)
- Plextor 716UF (USB/FireWire)
- Plextor 755UF (USB/FireWire)
- Plextor PX755 (SATA)
- Plextor PX760 (IDE)
- Plextor Premium (IDE)
- Plextor Premium 2 (USB)
- Plextor Premium-U (USB)

CD Error Levels

Error checking with a Plextor drive includes the C1/C2/CU error test. C1 errors are present on every disc, C2 errors are more serious errors but are unlikely to affect playback, and CU errors (also called *BLER errors*) are serious and will affect playback. The Plextools error test and the printed test results are shown in Figure 12-1. The *Red Book* standard requires a maximum of these error levels:

- C1: Average of less than 220 errors per second
- C2: No C2 errors
- CU: No C3 errors

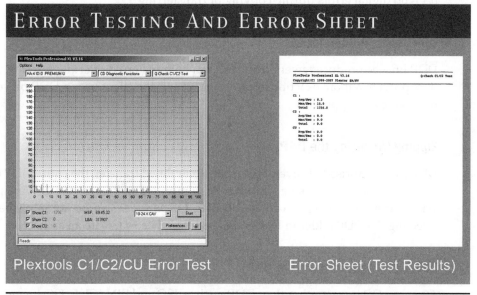

FIGURE 12-1   Plextools error testing is the standard in audio mastering studios. This figure shows the test screen and the printed test results that are provided when delivering the premaster CD.

CD manufacturing plants usually have their own error limits and are provided with error test sheets by mastering studios. At Stonebridge Mastering, we ensure that C1 levels are less than 3.0, with no C2 or CU errors. We stamp these error sheets with a customized rubber stamp as certified premaster CDs.

DDP File

Disc Description Protocol (DDP) files are CD images from which a glass master CD can be created. DDPs can be created by a mastering studio and sent over the Internet to a duplication/replication plant. This avoids shipping costs and possible problems associated with premaster CDs. Several mastering DAWs, including Magix Sequoia, will create DDPs within the DAW. There is also stand-alone DDP creation software such as Sonoris DDP Creator that performs this task.

When DDPs are created and CD-Text will be used, the language should be set to the language of the CD (e.g., English, Spanish, etc.). Leaving it undefined can cause red flags in CD replication software. For example, the CD manufacturing software Eclipse will give an error saying, "CD-Text bad language code." Also, nonstandard characters in CD-Text such as exclamation points can cause problems.

DDP files contain four main files: image.DAT, ddpid, ddpms, and sd. The image.DAT file contains the audio for the CD image. The ddpid, ddms, and sd files contain all other information related to the CD (the PQ codes and CD-Text). DDP files also may contain an MD5 checksum that you can use to be sure that there has been no digital errors during copying or transferring.

DDP versus Premaster CD

DDP images produce fewer errors and can be transferred over the Internet. Also, they provide a more exacting approval process if the client uses a DDP player for reviewing and approvals.

Zipping/Archiving the DDP

DDPs are comprised of several files. For streamlined transfer, the entire DDP file is usually "zipped" into a single file. For PC computers, software such as Winzip and Winrar can be used for creating Zip files. On Macs, it can be accomplished by selecting the DDP folder in Finder and selecting File → Create an archive.

BIN/CUE (An Alternative to DDP)

There are CD images other than DDP, such as ISO and BIN/CUE. ISO images only work for data CDs and cannot be used for audio CDs. However, the BIN/CUE image format does work for audio CDs. There are a variety of CD-burning programs that will work with BIN/CUE files on both PC and Mac, including Toast, Roxio, Nero, MagicISO, Alcohol 120%, ImgBurn, LiquidCD, and many others. There are even a few CD-pressing plants that can work with BIN/CUE files in addition to DDP. There is also WAV/CUE, which is nearly the same as BIN/CUE. However, with BIN/CUE and WAV/CUE files, some burning software will not properly burn metadata such as artist information and ISRCs. It is thought of as being somewhat less than totally reliable.

Other DDP Alternatives

Exact Audio Copy and Burrrn are free programs that have some of the functionality of DDP.

Drawbacks to DDP Alternatives

Almost every replication plant will at some point have to convert to DDP files to perform their tasks, so sending them a DDP saves a step. Many will not accept any alternative except a premaster CD. Also, the alternatives are typically not as reliable.

Mastering for Vinyl

If a vinyl mastering engineer will be working with the project, often he or she is provided with a mastered version that does not have final limiting. This makes the cutting process easier. Also, careful attention is paid to bass levels. Bass frequencies might be removed from the side channel to prevent too much vertical movement of the needle, especially if loudness is a priority. There is more information on this in Chapter 7. Although very rare, there are some recordings that work well with panned bass as an intentional effect, which is especially popular with some jazz recordings. With these kinds of recordings, where removing bass from the side channel is not an option, the side channel can be limited, or the record can be cut at a low level. Also, when preparing for vinyl, adding "air" using equalization is best kept to a minimum. For more on this topic, check out Jeff Powell's "Premastering for Vinyl" in Chapter 16.

Shipping to the Client

Mastering studios often provide clients with two masters so that both may be provided to the CD manufacturer. If there is a problem with the first CD, the second CD can be used without needing to hassle the client. Clients are normally provided with PQ sheets and an error sheet for each CD, which are forwarded to the CD manufacturer. Providing instructions to the client about these items is a good business practice.

PQ Sheets

It is common knowledge that CDs have tiny ones and zeros, represented by microscopic *pits* and *lands*. Those tiny ones and zeros are organized into small units of data, the smallest of which is called a *frame*. A CD frame has various parts, including the audio data itself, parity, syncing data, and something called *subcode*. The subcodes are 8 bits long and are labeled P, Q, R, S, T, U, V, and W. The P and Q bits are the only ones that are used with digital audio CDs. These are where the terms *PQ codes* and *PQ sheets* were derived.

The P bit contains a value that indicates whether its frame is the start of a new track, the very start of the CD, or the very end of the CD. The P bit also designates the pause time length.

The Q bit contains information about number of audio channels (two or four), the number and start times of tracks, and CD-Text information.

PQ sheets show the P and Q subcode information—the start and end times, pause times, total running time, and CD-Text. There are a few different styles of PQ sheets, including the Magix style and the Sonic Solutions style. The Sonic Solutions style is often best because it will work with longer names. It is possible to switch between these styles in the newest versions of Samplitude and Sequoia. Both styles of PQ sheet are shown in Figure 12-2.

When transferring a DDP to a CD manufacturer, the PQ sheet is most often provided in PDF format, separate from the DDP.

Shipping to the Replicator/Duplicator

CD manufacturers can be sent a physical premaster CD, PQ sheet, and error sheet, or they can be sent a DDP image and a PQ sheet (often in PDF format) over the Internet.

Checksum/MD5

There are many free programs out there that will perform checksum testing. The most popular checksums for mastering engineers is the Message-Digest Algorithm 5 (MD5). This testing verifies that the digital copies are exact. MD5 codes can be generated from a digital file and e-mailed. Then a MD5 verification application can be used to make sure that the digital file received is an exact copy of the file that was sent. This eliminates the possibility of transfer errors.

PQ SHEETS

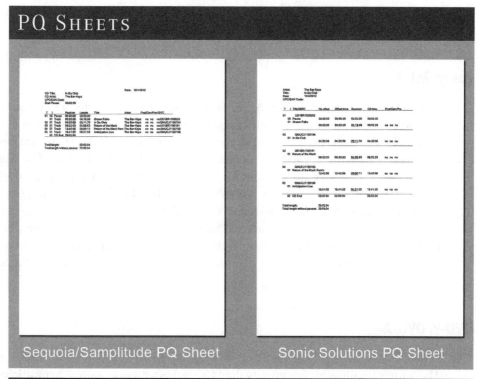

Sequoia/Samplitude PQ Sheet   Sonic Solutions PQ Sheet

FIGURE 12-2   These two PQ sheet styles show information about the premaster CD, including track times, ISRCs, artist name, and track titles.

Manually performing this testing is quite rare. DDP files automatically contain MD5 checksums as a protection mechanism. It is part of the format.

Nonlossy: WAV, AIFF, FLAC, Etc.

Some clients like to have nonlossy versions of their recordings. WAV files are the most common nonlossy audio format, likely followed by Audio Interchange File Format (AIFF). Free Lossless Audio Codec (FLAC) files have been increasing in popularity and are usually about half the size of a WAV with no loss of quality.

Nonlossy Metadata

Metadata are descriptive information embedded into a digital file. Each of the nonlossy formats listed above is capable of having metadata embedded, with the exception of WAV files. However, there is a format called *Broadcast WAV* (BWAV) that does allow metadata to be embedded. Also, for embedding metadata into

WAV files, there is the Cart Chunk WAV format, ratified as the AES Standard AES46-2002 (for more information visit cartchunk.org).

Lossy: MP3

When the final medium will be MP3, there are several techniques for optimization. Professional MP3 files are usually encoded from 24-bit or 32-bit sources because they provide more accurate conversion than 16-bit sources. There are several different MP3 codecs used for encoding audio into MP3 format. The Sonnox Fraunhofer Pro-Codec plug-in allows an engineer to audition various codecs to discover the one that best suits the recording at hand. When mastering for MP3, the limiter ceiling is often lowered to −1 dBFS. Also, rolling off the highest frequencies while monitoring and comparing with the Sonnox Fraunhofer Pro-Codec can allow for higher quality. Higher frequencies are affected the most during MP3 encoding, and rolling them off can produce a gain in quality across the rest of the spectrum.

DVD-V, DVD-A

When DVDs are used for audio, it is with either the DVD-V or DVD-A format. These formats allow higher fidelity than CD and offer mono, stereo, or multichannel audio support (up to 5.1 surround). Both DVD-V and DVD-A have a bit rate of 24 bits. While DVD-V has a maximum sample rate of 96 kHz, DVD-A can go up to 192 kHz. DVD-A also has more storage capacity for audio than DVD-V. Most DVD players being sold today are capable of playing DVD-A and have been for years. Software including Magix Samplitude and Sequoia is capable of DVD-V and DVD-A authoring.

SACD

Super Audio Compact Disc (SACD) is a 1-bit audio CD intended to give the listener a high-fidelity audio experience. The SACD format requires the consumer to buy a special SACD-compatible player. Also, SACDs and high-quality SACD players are rare. SACD does offer a high-quality recording, but there is no multichannel support, only mono and stereo. There are no SACD manufacturers in the United States, and releasing in this format is very expensive.

Blu-ray Audio

The Blu-ray Disc (BD) is likely to be the last physical disc medium to be manufactured. Also, many consumers own Blu-ray players and do not have to buy special equipment, as with SACD. Blu-ray can be used for audio and can do everything that DVD-A can do, except with more storage capacity. The drawback with Blu-ray audio is the high expense of equipment and software for manufacturing and authoring.

Mastered for iTunes

Special "mastered for iTunes" recordings are being sold through Apple iTunes. Mastering for iTunes is the optimization of a recording for the Advanced Audio Coding (AAC) format (the format used by iTunes).

Apple has published a list of guidelines for mastering engineers to follow for optimizing a recording for conversion to its AAC format. When mastering for iTunes, recordings should not be converted to 16 bits but instead left at 24 or 32 bits. Also, there is no benefit from upconverting or upsampling. No sample-rate conversion or dithering should be applied. Apple recommends auditioning the masters on the current and most popular consumer listening devices, such as its white earbuds, other popular headphones, and speaker docks. A limiting ceiling of –1 dBFS is very strongly suggested but not mandatory. The guidelines mention Apple's new feature called *Soundcheck* that adjusts the listening volume based on RMS/average levels. This means that when mastering for iTunes, the optimal dynamic range should be used because there is no benefit from creating louder masters if the user has the Soundcheck feature enabled.

Although Apple does not mention it, the Sonnox Fraunhofer Pro-Codec software plug-in allows the AAC format to be auditioned while listening, so the mastering adjustments can be made while hearing how the AAC conversion will affect the sound.

Clients are normally provided WAV files that are ready to be encoded to AAC format. Their digital distributor normally will perform the conversion to AAC with the iTunes Producer application, which is used to submit information and recordings into the iTunes store.

Enhanced CD

There are a number of applications that allow for creating enhanced CDs. One of the most common is Sonoris DDP creator, which allows enhanced CD content to be added to a DDP file under Tools → Create ISO.

Ringtones

When mastering for ringtones, there are a few common techniques. It is often best to roll off the lowest lows. Limiter ceilings should be set lower than normal, a ceiling of –0.8 dBFS is more typical. Also, less limiting is used to minimize distortion of the small phone speakers. Because most ringtone playback will occur in mono, it is important to monitor in mono while making adjustments.

5.1 Audio

If a client requests 5.1 mastering and the output is not requested to be on DVD-A, DVD-V, or Blu-ray, then stems are normally provided without any conversion, sample-rate conversion, dithering, or bit-rate conversion.

CHAPTER 13

What Happens After Mastering?

After mastering is complete, the journey of the album is just beginning. There are a few details to take care of, and soon the recordings will be heard on a variety of listening systems. Sometimes, at this point, mastering clients need a little guidance about what to do next. Also, there are things that happen after mastering that may affect the way you work. This chapter provides a view of what happens after the mastering role is complete.

Storage and Returns

Often, if you are working with analog tape, the tape will be returned to the record label for storage. If analog tapes are to be stored by the mastering engineer, they are usually stored in plastic bags along with desiccant material (material that absorbs moisture). Digital media used to transfer files normally are kept by the mastering engineer and stored, along with any information about a project. External hard drives and online cloud storage provide excellent means to store digital files and protect against loss. Also, it is important for clients to maintain a copy of final mixes and not to send the only copy to mastering.

Database Submission

Some CD players show CD-Text, but most software music players and services draw information from online databases. For example, iTunes draws from the Gracenote database. Windows Media Player and the Playstation 3 player draw from the AMG Lasso database. The AMG Lasso database requires a physical copy of the CD be sent to it, from which it enters the information. AMG Lasso is infamous for incorrectly entering information, so it is recommended to type out a sheet to include with the submission with song titles and other information in a large, easy-to-read font. Below is a list of the most common databases:

- All Media Guide/AMG Lasso
- Gracenote
- Muze
- FreeDB
- MusicBrainz

AllMusic Credits

The most widely recognized clearinghouse for album credit information is the AllMusic Database. Most often the client's digital distributor will be making the AllMusic Database entry. It's often difficult to make sure that the mastering credit is added. Communication with the client to advise him or her about requesting the credit can help. It is also possible to communicate with AllMusic but it can be difficult to arrange for additions or changes.

Replication versus Duplication

CD copies are made by a manufacturer using one of two processes—replication or duplication. With replication, a glass master is made. A glass master is created by covering a block of glass with a chemical coating that contains the data of a CD. Then it's used to stamp out the CD copies. In the most precise language, this is called *mastering*, and the service performed by mastering engineers is called *premastering*. The duplication process is entirely different from replication. In duplication, a CD is copied using the CD-burning method familiar to most consumers. Usually, CD duplication machines allow multiple copies to be made at once to speed up the process. Replication is normally chosen for high-quantity orders (1,000+), whereas duplication is for smaller quantities. During this process, the manufacturer also prints the CD artwork, and the final salable products are assembled and shrink wrapped.

Eclipse Systems

Many replicators use Eclipse systems (by Eclipse Data Technologies) to create glass masters for replication. The "imagecopy read-in" feature of Eclipse analyzes masters for errors. Often problems with CD-Text, including non-standard characters or the language set to "not defined" in a DDP image, can produce errors. If errors occur during this stage, replicators usually notify the client that an error has occurred in premastering. Even when the error returned by Eclipse is in line with the *Red Book* CD standard and is not a true error, most mastering engineers wish to avoid being perceived as having made an error.

Disc Life

The strictest experts say that a burned CD may have a life of only two to five years; manufacturers say that they may last from 50 to 200 years. The quality of the disc is a big factor in disc life, and the life varies among brands. Gold CD-Rs provide a longer-term option and are used for archival material. They are said by manufacturers to have an expected life of around 300 years, although this remains to be seen.

Radio and Broadcast Processing

In FM/AM radio processing, Orban and Omnia are the two most popular manufacturers. Omnia broadcast processors are the most common in radio stations across the United States. The most noticeable processing that takes place with these processors is multiband limiting (called *multiband AGC* in radio terminology). These processors have sophisticated loudness-detection algorithms, so all recordings they process are played at equal loudness. Because of this, creating loud masters will not make a song sound louder on the radio. Radio processors also have equalization processing that can affect recordings, especially the bass frequencies. To find out more, see Cornelius Gould's "Optimizing Audio for Radio" in Chapter 16.

Car Stereo

Because of the positioning of the speakers, the listening environment of a car produces a sound that tends more toward mono than stereo. Also, car stereos have adjustable equalizers, which are typically shelving equalizers.

CD Refractive Index

Many people have had the experience of burning a CD only to have it fail to play in certain sensitive CD players. Mastering engineers who are just starting out may have performed CD error checking and noticed that the error levels are far below the *Red Book* standard error rates, yet the CD still will not play. The issue at play here is likely to be the CD refractive index levels. Silver CDs produced from a glass master have extremely high reflectiveness, so these CDs will play in sensitive players. Burned CD-Rs, of any type or error level, may not work with the most sensitive players that require the highest in CD refractive index levels.

Clubs

In dance clubs, the acoustics and PA system are normally such that the very lowest bass sounds do not come through powerfully. Bass sounds that have the most energy in these environments are normally between 75 and 95 Hz.

Growing Home Theater 5.1 Systems

Because of the increasingly low cost of equipment, consumers are buying 5.1 home theater systems. As the sales of 5.1 systems grow, it presents an opportunity for studios to produce in this format.

Internet/Streaming/Format Conversions

Mastered audio often finds its way into many different audio and video streaming formats on and off the Internet. When these conversions occur, if the limiting ceiling is set too high, there can be clipping problems.

Smart Phone/MP3 Player/Computer Playback

Smart phones, MP3 players, and random song playlists are the one area where the loudness of a recording can make a difference. When a mastered recording is played in a playlist with audio from other albums, the levels will not be consistent. Because of this, the impact to the listener may be lower with recordings of a lower loudness.

CHAPTER 14

DAW/Computer Optimization and Interfacing

oday, every professional mastering studio has either a PC or Mac. PCs are a bit more common because Magix Samplitude and Magix Seqoia, which are very popular for mastering, work on the PC exclusively. This chapter only covers the PC.

When your computer slows down or has problems, it can wreak havoc on your work. This chapter covers the main topics that are important for mastering engineers to keep a PC running smoothly. When in doubt, contact an information technology (IT) specialist.

In this chapter, when discussing operating system tweaks, the examples refer to Windows XP. If you are using a different operating system, it may be necessary to search Google to find out how to perform the tweaks, although they are generally the same in all recent versions of Windows.

Do Not Let a DAW Dictate the Workflow

Just because a digital audio workstation (DAW), computer application, plug-in, or device has an implied workflow, it doesnt mean you must follow it. Be mindful of the fundamentals.

Save Early, Save Often

When something good happens, save. If a miracle happens, save an extra version. If a plug-in chain is vital, save the plug-in chain/settings separately in case an unfreezing fails to work. Save often, and save yourself from headaches.

Keyboard Shortcuts/Hotkeys

Using keyboard shortcuts, also called *hotkeys*, can increase productivity. For example, zooming operations are very well suited for keyboard shortcuts.

DAW Functions

Understanding the features of your chosen mastering DAW is vital. A few of the most important DAW functions in mastering include cross-fades, track volume automation, markers, CD-Text, International Standard Recording Codes (ISRCs), plug-ins/processing, routing/routing sequences, solo modes, and snapping functions. These are DAW-specific, so using these features must be discovered from the DAW user manual or other resources.

Wacom Tablets/Trackballs/Mice

The hardware you use to work with your computer is more important than most people would think. The range of interface devices used while working on a computer is worth exploring for mastering engineers. For some users, Wacom or similar tablets speed up workflow tremendously and allow for a more tactile experience. Others prefer trackballs or "gaming" mice. It is also worth considering USB foot pedals to replace common keyboard shortcuts. With many DAWs, it is possible to assign an operation to a keyboard shortcut even if one does not exist in the default setup.

RAID and Online Backup

Essentially, redundant array of independent disks (RAID) systems consist of two hard drives that function as one, each being an exact copy of the other. RAID systems are as fast as having one drive and maintain a constant backup. This redundancy protects against losing data during a physical hard-drive problem.

Services such as Mozy Online Backup, Acronis, Crashplan, DropBox, and others maintain online backups. This cloud backup can be slow for mastering studios because audio files are very large. Such online backup options require a fast Internet connection.

Solid-State Drives (SSDs) versus Hard-Disk Drives (HDDs)

Modern solid-state drives (SSDs) provide a performance boost, although standard disk drives are more cost-efficient. At this time, for mastering engineers, the performance boost does not significantly boost productivity.

Ending Nonessential Processing and Services (Windows)

With Windows systems, Task Manager is often used to view the processor and memory load, as well as for closing application and process instances. Sysinternals Process Explorer is a more advanced version of Task Manager and is a great tool for discovering more information about a system's active processes. A wealth of information about processes can be found using the View → System Information option of Sysinternals Process Explorer and mousing over the activity in I/O. When a spike is seen, mousing over it in Sysinternals Process Explorer displays the services causing the spike. This can help in discovering the root of the problem. Working to solve these kinds of problems must be done carefully so as to not cause operating system problems. Sysinternals Process Explorer is shown in Figure 14-1.

User Accounts

Windows has a feature where multiple *user accounts* can be used. Sometimes it is recommended for mastering studios to set up a separate user account for mastering. Windows services can be assigned to various user accounts. The user account used for mastering is often set up so that only a bare minimum of services is enabled. Often services used for networking and Internet are disabled for this account. Audio applications may be set up to run on all user accounts so that they may be updated via the Internet with an administrator account. Setting up processes to run under user accounts involves navigating to Administrative Tools → Services under Windows XP or executing services.msc using the Run dialog box. Services can be right-clicked, and under Properties → Log-on the services can be set up for different user accounts.

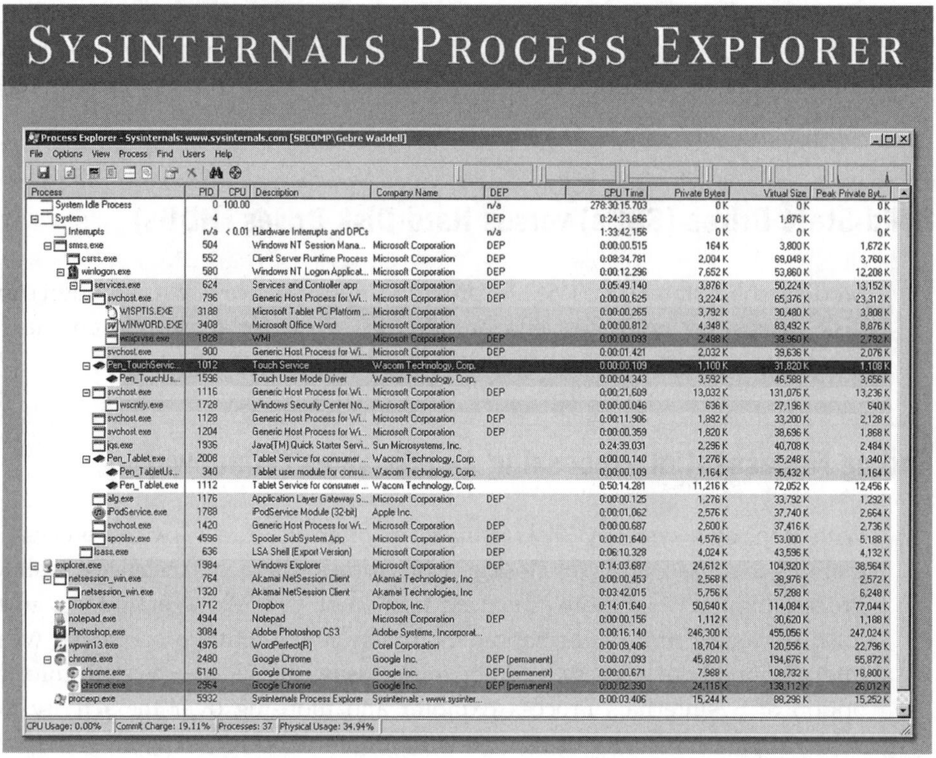

Figure 14-1 Sysinternals Process Explorer is an application that can help to find the source of many problems on Windows-based systems.

Warning *Setting up user accounts and services in this fashion incorrectly can cause severe operating system problems. This setup usually requires the assistance of an IT professional.*

Faulty Drivers

Drivers are software programs that help hardware devices communicate with a computer. Software drivers sometimes have bugs that cause problems such as crashing or dropouts when working with audio. Some bugs require the manufacturer to program a new version of the drivers. Usually problems associated with drivers are caused by conflicts with other drivers or devices. Sometimes disconnecting things or turning things off one by one can help in finding problems with conflicts. Interrupt Resource Request (IRQ) conflicts are one of the most prevalent. Opening Windows Device Manager, navigating to View → Resources by Type, and expanding the IRQ tree, will display hardware devices by their IRQ usage. Nonmotherboard hardware occupying the same IRQ could be a potential problem. Again, this usually requires the assistance of an IT professional.

Plug-in/Interface/DAW Conflicts

Plug-ins may not be perfectly compatible with every DAW. There are variances in adherence to the VST and Direct X standards. Even some audio interfaces do not work well with some DAWs. If these kinds of conflicts are suspected, swapping things out can isolate the problem.

Other Typical Problems/Regular Maintenance

Fragmentation, hardware failures, IRQ conflicts, and many other problems are possible when dealing with computers. A weekly or monthly maintenance schedule can help. The schedule might include scanning for problems and hard-drive defragmentation. Time-consuming scans can be set up to occur overnight.

DPC Latency Checker

The Deferred Procedure Call (DPC) Latency Checker application can tell you if a computer will be able to function as a DAW without dropouts. It also can help to troubleshoot problems. When red spikes are observed, there are many possible causes, including IRQ conflicts, other hardware conflicts, malfunction, system problems, driver problems, and spyware/badware/malware/viruses. It is possible to disconnect devices or disable suspected devices in Device Manager one by one until the culprit is isolated and the red spikes stop.

Computer Hardware Problems

Faulty random-access memory (RAM), faulty processors, and overheating are a few of the most common hardware problems. With these types of problems, you may experience random blue-screen crashes or spontaneous rebooting. RAM can be checked using a program called Memtest 86. If the processor is the problem, it can be checked with a program called Prime95, which forces the processor to make complex mathematical calculations and checks the results for errors. If there are errors, disabling multiple cores in BIOS setup can help. If this does not work, the processor is likely to need replacing, although processor failure is quite rare. Overheating is another common hardware problem and can be checked with a program called hmonitor. Hmonitor indicates the zone where overheating has occurred. Zone information is shown in the owner's manual of a computer's motherboard, which usually can be found with a quick Internet search. Fans can be added in overheating zones.

Operating System Tweaks

There are two main tweaks for optimizing a system for audio. The first is to turn off visual themes. It doesn't look so great, but there will be more memory for audio. The second is to set the central processing unit (CPU) priority for background services. This allocates more memory for the way DAWs work. In Windows XP, this can be accomplished by going to the Windows Control Panel and navigating to System → Advanced → Performance → Settings → Advanced → Processor Scheduling → Background Services.

Disable Onboard Sound in BIOS

Many motherboards include an onboard sound card. This sound card should be disabled in BIOS to ensure that there are no conflicts with the audio interface. On most Windows systems, the boot screen displays a key that can be pressed to enter BIOS setup. When this is not displayed, the computer or motherboard's owner's manual should be consulted about how to get into BIOS.

Disable Internet and Antivirus

Professional studios often disable their Internet connections and even their antivirus software on the computers used for mastering. This ensures minimal conflicts and the highest in stability and productivity.

Second Hard Drive for Audio

Often audio professionals use two separate hard-drives—one for the operating system and DAW and the other for digital audio files. Keeping these separate can help to optimize DAW operation.

FireWire

If a FireWire audio interface is being used, it is usually best to go into network connections and disable the 1394 connection. This connection can be uninstalled if

FireWire will not be used to connect to the Internet. Institute of Electrical and Electronics Engineers (IEEE) 1394 is another name for FireWire, and there can be problems with a computer attempting to connect to the Internet via an audio interface.

Startup

If undesirable programs or services start when the computer is booted, they can be configured not to do so. Go to Run, and type in "msconfig." A utility will open that will allow configuration of startup options.

System Registry

The Windows System Registry is a database of system options and configurations. System registry problems can cause errors that slow things down and cause crashes and other problems. A Google search of "best registry scanners" or a similar search term will provide options for applications that can scan the system registry for problems and repair them.

Latency/Buffers

Audio buffer levels usually can be set both in the interface software and the DAW. These options are more important during tracking and mixing but can be a factor in mastering. Increasing the buffer size creates more system stability, although it increases the audio latency during recording.

Driver Systems

Most DAW platforms provide a choice of whether to use Audio Stream Input/ Output (ASIO) or Windows Driver Model (WDM) driver systems. The choice between these two driver systems can change the sound quality, latency, stability, and input options. In most cases, ASIO is best, although it can depend on the audio interface and desired options. To find out more, see David A. Hoatson's "ASIO versus WDM Drivers" in Chapter 16.

Spyware/Badware/Viruses/Malware

If malicious software such as viruses, adware, spyware, or rootkits are causing a problem, there are a few tools out there to help. It would be impossible to list all the applications that repair these problems, and each may find things that the others miss. Updating and running every application listed here will remove nearly all threats. AVG Antivirus has a free version of its virus scan software available. Malwarebytes' Anti-Malware is great for getting rid of all kinds of issues and is also free. SuperAntiSpyware Free Edition may find things that the others do not. Spybot Search and Destroy is another free program that can fix a wide range of problems. ComboFix is an especially powerful free program that can help to remove rootkits, which are similar to viruses. If everything else has failed, try Norton Power Eraser—it can sometimes cause problems, but it is very effective on the toughest problems. Finally, Hijack This is one of the best programs out there for spyware/malware removal, but it requires carefulness and IT experience to use it without causing problems.

AVG Antivirus	http://free.avg.com/
Malwarebytes' Anti-Malware	www.malwarebytes.org/mbam.php
SuperAntiSpyware Free Edition	www.superantispyware.com/
Spybot Search and Destroy	www.safer-networking.org/index2.html
ComboFix	www.combofix.org/download.php
Norton's Power Eraser	http://us.norton.com/support/DIY/
Hijack This	http://free.antivirus.com/hijackthis/

Be Serious About Technical Problems

If computer problems become a burden, it is important to think seriously and creatively to solve them quickly. The information here, combined with research on Google, can help. Computer stability is a valuable foundation of mastering.

CHAPTER 15

Starting a Mastering Studio as a Business

For those of you who love audio mastering, at some point you must decide whether you want to make it into a career. There's your love for the art, and then there's the risk of starting a business. You face competition, considerable startup costs, and many other challenges. This chapter focuses on the things you must consider while making this important decision. I'll start with the tough challenges that new studios must face and end on a high note with the benefits of opening a new mastering studio. I won't be covering luck, although a bit of it is needed as well.

Challenges Faced by New Mastering Studios

For anyone considering opening a mastering studio, a careful decision lies ahead. It's something that takes serious time and investment. As with anything, it requires sacrifices that can change the course of your life. Also, there are several challenges faced by new mastering studios.

Entrenched Competitors

Competition is something every entrepreneur must consider, and with mastering, there are a few special considerations. The established studios have entrenched relationships with the major labels. The most well-known mastering studios, including Sterling Sound, Masterdisk, Gateway Mastering, and Bernie Grundman Mastering, among others, have spent decades consistently proving their abilities and honing the skills of mastering. Such studios are focusing even more on independent artists, making it possible for anyone to obtain their services. Also, many engineers who work for the biggest studios leave and start their own separate, highly respected studios. In mastering, the competition is heavy and well entrenched.

Do-It-Yourself (DIY) Mastering

New mastering studios market their services to independents and home/project studios. With mastering software such as iZotope's Ozone, which contains decent tools at a low price, customers that otherwise would visit a new studio with a low price often choose to spend their money on software to perform their own low-budget mastering. Low-budget tools are accessible by a group that is very do-it-yourself-minded to begin with.

Customer-Service Challenges

Perhaps the most unforeseeable challenge for new mastering studios is the customers. During the first 10 years in business, the challenge of customer service is intense. Despite the information out there, clients usually do not know the basics of mastering. Many clients expect more than what's possible in mastering, and when they are presented with their masters, they will have additional, often impossible requests. Charging more for additional requests reduces competitiveness. Customer satisfaction is often low as new mastering studios work toward honing their abilities and dealing with bad mixes. As one Chicago mastering engineer once put it, "I had to ruin a lot of albums before getting to this point," and surely many albums were ruined before they ever got to him.

False Advertising and Scams

Widespread false advertising and outright scams are a big issue in mastering right now. It is commonplace for new mastering studios to falsify their gear lists, claim fake awards, and make false claims about working with major labels. They launch a stellar website, list major labels, sometimes add some stock photos (or no photos), and they are in business. In large part this may be due to the difficulty of starting a legitimate studio. This climate has led to a distaste for mastering because of the misleading experiences many customers have had. After a disappointment, dissatisfied customers usually seek out one of the big studios or just perform DIY mastering themselves.

Album Sales at All-Time Low

Another significant challenge is that albums are selling less and less. Today's climate for the recording industry is tough—only 1,215 albums sold over 10,000 units in the United States in 2010.

Nonscalable

Mastering as a business model is not scalable. This makes it uninteresting to the vast majority of entrepreneurs. Mastering studios are usually started by a single individual who enjoys mastering or companies somehow related to the field. New mastering studio owners usually have some existing connection to the music industry that will afford them opportunities while getting started. If business is your passion, something scalable is more likely to provide success than starting a mastering studio.

Low Growth Rate

The projected growth rate is a consideration when entrepreneurs evaluate a business model. Few would disagree that the growth rate in this industry is low and that the business cycle for mastering is in a plateau stage that is likely to remain unless a technological shift occurs. This plateau has been reached after some years of declining from the peak.

High Startup Cost

The startup cost of a truly competitive mastering studio is relatively high because the acoustic environment is of prime importance and requires costly structural modification or specialized architecture. Also, the cost of analog processing equipment is significant. Although a few less-expensive digital plug-ins are truly mastering-grade, analog equipment still offers many highly demanded, unequaled strengths.

No Points

Unlike some positions in music production, mastering engineers almost never receive percentage points on album sales.

Limited Scope

Mastering as a service is usually respected only if it is the sole concentration of a business. Therefore, having a successful mastering business usually means being limited in scope.

Legal Disclaimers/Limited Liability

In almost every area of business, liability should be limited for possible damages. The biggest risk in mastering is that a large quantity of discs may be pressed that contain an error. Quality control is key in preventing this problem. Also, a second level of protection is provided if a mastering studio requires its clients to check the final medium before sending it for replication/duplication. The MD5 checksum that is possible with DDP exporting is a great way to limit the liability if an error is introduced at the manufacturing stage. Keep in mind that there is no substitute for consulting a lawyer to discuss limiting liability with any kind of business.

Benefits of Starting a Mastering Studio

While the challenges are tough, there are several benefits to starting a mastering studio.

Talent in Practice

Possibly the greatest benefit to starting a mastering studio is for those who are naturally talented in mastering. As the practice of law or medicine might allow someone to express his or her natural talent for those professions, mastering studios afford individuals the opportunity to practice their talent for mastering.

Marketing Without Limitation

One of the greatest benefits of starting a mastering studio is the ability to market worldwide. Audio recording and mixing services usually require more subjectivity and the physical presence of everyone involved. Because of this, recording and mixing services are more suited for local markets. Far too often, new tracking and mixing engineers overestimate their ability to compete with established studios in their local markets. With the low cost of recording and mixing gear, the challenge is tremendous. Instead, mastering is less subjective than tracking and mixing and may be best without the client present.

Surge of Independent Artists

With the rise of independent music, there are more people seeking mastering services for the first time.

Shorter Projects

Compared with audio tracking and mixing projects, which can stretch from weeks to years, mastering projects are very short term, lasting for hours and, at most, weeks.

Less Band Politics

Tracking and mixing sessions can include a deluge of requests and issues. With mastering, there is usually less to discuss.

Local Attended Sessions

If a studio is opened in an area where there are no local mastering studios, the local market may provide an opportunity. Some local clients may prefer attended sessions and would like to avoid travel.

Being a Sole Proprietor

Because mastering engineers are frequently sole proprietors, there is a freedom in working for yourself, especially for individuals who function better independently than in groups.

Rising to the Challenge

Who is best suited for rising to the challenge of a career in audio mastering? The best mastering engineers are patient, great at communicating with clients, naturally dedicated to audio engineering, and willing to narrow their scope to mastering. While rising to the challenge of starting a mastering studio isn't for everyone, there are some things that can help along the way.

Seek an Internship

One of the best ways to get into mastering is through an internship with a respected engineer. Such an internship is a perfect opportunity for understanding the process and can help to jump-start a career. Internships should be taken very seriously.

Take a Course

Berklee College of Music in Boston offers an online audio mastering course called Audio Mastering Techniques.

Collect Resources

Read books, forums, and blogs. Watch videos and listen to podcasts. In Appendix B you will find a comprehensive list of mastering resources.

Learn Aggressively

Dedicate yourself to learning. Maintain a collection of all the mastering resources, explore everything, take the course, meet people, practice the techniques, continuously study this book, and do everything within your power to learn.

Learn About the Legends

There are many people who have devoted their lives to this work. Learning about them provides valuable lessons.

Add Mastering Services to an Existing Studio

Existing recording studios may be open to starting a mastering studio that is attached to their brand. This can be a great way to get started with a steady customer base. The prestige of the existing studio will be passed to you, and some customers will still seek your services in the event that you separate from the studio.

Leverage Existing Opportunities

It is important to think creatively about opportunities. If you live in a cultural center such as New York or Los Angeles, then you may be in close proximity to great studios and could more easily obtain an internship with a legendary engineer. If you live in a city that has a flourishing music culture but does not have a local mastering studio, you may be in luck. If you have family members or friends who are in the music industry, they may be able to help obtain notable work, improving the impact of your client list.

Reach New Markets

You must find a way to reach new markets. In mastering, word of mouth is very important, so doing great work for someone in a new market may be the very best way to connect.

Visit the National Association of Music Merchants (NAMM), Join the Audio Engineering Society (AES) and Grammy Recording Academy

Joining the AES and Grammy Recording Academy is a great way to get started. Attending NAMM, the AES Convention, and your local Grammy chapter meetings can help to network nationally and internationally.

Working for Notable Acts

People often equate the skill of an engineer with the popularity of his or her previous clients. This means that it is important to have a strategy for getting notable work. These kinds of opportunities are usually unique to the individual.

Insure Your Studio

Anything can happen. Some insurance companies such as Capital-Bauer and MusicPro Insurance specialize in studio insurance.

Hiring Interns

Some mastering studios receive the help of interns from local universities. Some mastering studios find the challenge of working with interns to outweigh the benefit. Internship programs may work best when training is optimized by a system where interns train interns; however, a string of unmotivated individuals can halt the system.

Other Opportunities

There are opportunities related to mastering that one also might take into account while considering a new mastering studio.

Opportunities for Audio Forensics

Some engineers, when starting out in mastering, plan on providing audio forensic or other audio services to legal professionals. Generally, if there is audio evidence in a legal case, it cannot be presented to a jury in an altered way without the testimony of an expert. Legal requirements for someone being an expert are usually quite high, and the person's qualifications will be thoroughly scrutinized and subjected to cross-examination. Today, the New York Institute of Forensic Audio is the only certifying agency for such experts.

Consider Related Fields

When examining this area for career opportunities, you also might consider working with a business-to-business model by serving a market of audio mastering engineers. Digital signal processing (DSP) programming/algorithm design and analog circuit design or repair are two areas that may provide more stability and career options than providing audio mastering services. It's all about personality and interests.

CHAPTER 16
Contributions

This chapter features contributions to this book by some of the most respected individuals in mastering and related fields today. Each contributor provides a unique and valuable perspective on their topic.

Robin Schmidt is the owner of 24-96 Mastering in Karlsruhe, Germany. His work has been popular throughout Europe and the United States, including releases with artists such as Two Door Cinema Club, OMD, and The Black Keys. His work on Louis Spohr's "String Sextet" was nominated for a Grammy Award for Best Surround Album in 2011. His work also includes Jake Bugg's self-titled album (number 1 on the U.K. charts in 2012), Ben Howard's Mercury Prize–nominated platinum-selling debut album "Every Kingdom," and The Coral's "Butterfly House" (U.K. album of the year in 2010).

Robin Schmidt helps us to understand how he chooses equipment. He uses this topic as a base to explain part of his mastering philosophy.

On Analog Multiband Compression and Audio Gear

Robin Schmidt, Mastering Engineer
Owner of 24-96 Mastering, Karlsruhe, Germany

I bought an analog multiband compressor a few years ago on the recommendation of several mastering engineer friends. The package was very appealing: three bands of smooth opto compression, followed by a tube stage that, depending on how hard it's driven, was able to go from a very clean sound to delivering quite a bit of attitude.

After trying a few different sets of tubes, I settled for a combination that sounded clean and controlled until pushed hard, with low noise and microphonics.

I started using the unit on actual masters in the following weeks and liked its compression behavior and sound signature. But I found that even with great tubes and a well-calibrated machine, stereo levels fluctuated more than ideal. The reason for this is simple: In any analog compressor, there are parts that don't behave exactly linearly and have manufacturing tolerances. Opto cells, used for gain reduction, for example, never exactly match, and level controls (variable pots) are typically manufactured to allow a several-percent deviation.

With a normal stereo compressor, this means that the left/right balance may need adjustment after processing because the left and right level controls may be tracking slightly differently at any given setting. One channel may need to be turned up or down from indicated knob positions to match the other in real audio level, and a good mastering engineer will routinely check that left/right tracking is spot on.

However, with an analog multiband compressor, this gets more complicated. Consider that you now have potential level deviation between left and right channels in every frequency band. So, with any given setting, the bass frequencies may lean half a decibel to the left, the mids 1 dB to the right, and the highs may be exactly center. This isn't much of a problem in a recording situation, where amounts less than 1 dB are rarely that critical, but in mastering, these image shifts can become audible.

Luckily, the manufacturer was well aware of the requirements of mastering engineers and has a mastering version of the unit available. It employs resistor-network switches instead of pots and, in that way, gets rid of most of the stereo tracking deviation. So the normal unit was sold and a mastering version acquired instead. This one had much better stereo tracking, and the sound was a bit tighter to my ears as well. Perfect.

So there it was, the ideal analog opto/tube multiband compressor with the perfect tube combo, tightly calibrated. It sounded great, behaved well, and was a joy to use.

Still, over the months that followed, I found myself using it less and less.

The unit sounded great for imparting some tube tone or a slight overall sweetening, but when that was all that was needed, I tended to use other, simpler options, such as a vari-mu compressor or even just a set of transformers in the signal chain.

Instead, most of the time when I reached for multiband compression, I did so with specific tasks in mind—to control an uneven low end, to tame a screechy hi-hat, to soften sharp high mids in a vocal, etc.

After I had just gotten my analog multiband compressor, this was the default option for these tasks. But over time I found that more and more often I returned to solve such specific issues in the digital domain with multiband compression or dynamic equalizer plug-ins. While the analog unit sounded great, it lacked the flexibility, the ability to automate, and the ability to quickly compare different settings/approaches.

Analog multiband compression lacked the control that was so useful in dealing with these kinds of very specific, narrow, and inconsistent mix issues.

Sure, the unit sounded great. But what does that matter when you can't actually get down to the exact problem in the mix? Or when you can't apply it exactly in the ideal amount on only the relevant part? Or when the inability to compare different settings discourages you from trying different approaches to find the best way to solve the issue?

With a superb-sounding analog multiband compressor in the rack, I usually ended up choosing bog standard plug-in multiband compression instead so consistently so that in the end I sold the unit so that someone else could put it to better use.

This story, to me, illustrates a common quality that's often overlooked or not sufficiently mentioned when we talk about audio gear: that the most magnificent-sounding piece of gear is no good when you can't apply it exactly as you need to.

We often rave about the qualities, the sound color, or saturation characteristics of a Neve, Fairchild, Pultec, or other mythical box in absolute terms, as if they were black boxes, universal "bettermakers." But we rarely talk about how applicable a unit is, how much control the unit offers, what we can and can't do with it, and whether it encourages or discourages us from tweaking until it's just right.

I think it serves us well to acknowledge that in practice, when we're talking about manipulating audio, control *is* sound. The two are inseparable. Too little—or too much—control, and whatever sonic mojo a box might hold won't be used to its potential and might, in fact, limit the potential of the tweaking user as well.

The right amount of control is, of course, a very personal thing. A piece of gear that hinders one user's workflow might be perfect for another user. I know this to be true in the case of my analog multiband compressor. A mastering engineer friend of mine told me that he had to have his unit serviced and was dreading the downtime while it was out of the studio because the unit held such a vital place in his chain. For his workflow, the function he used the unit for, his taste, the unit had just the right amount of control.

And tastes don't just vary among different people, but also for one person, over time. Very often you'll find mastering engineers talking not about how they're currently gearing up, but rather about how they're gearing down, removing the "clutter" in their signal chains and workflow, essentially reducing the amount of control they have and feel that they are doing a better job because of it, because they find that they can better concentrate on the really relevant aspects of their work. A potentially cleaner signal path comes as an added bonus.

We all love our gear and have a very personal relationship with it. We select and collect it, invest in it, take pride in it, and love talking about it. Audio gear has heritage, it helps to create a shared studio culture, and it enriches our daily experience. It's a beautiful thing. But every now and then it's worth reminding ourselves that these pieces of gear are—above all else—tools, and as such, on the whole, they have to work for us rather than make us work for them.

Scott Hull is the owner and chief engineer of the iconic Masterdisk in Manhattan. Masterdisk contains several mastering studio suites and is one of the longest established mastering studios in the world today. Scott Hull's career began as an assistant for the legendary Bob Ludwig, who mastered classic albums of AC/DC, Rush, Jimi Hendrix, Madonna, Eric Clapton, David Bowie, Rolling Stones, Radiohead, and Nirvana, to name a short few. Hull's own career has developed to be one of the most noted success stories in professional audio mastering. He personally has over 1,600 album credits, including influential albums by Dave Matthews, Garbage, KISS, Miles Davis, Herbie Hancock, Bob Dylan, The Allman Brothers, Sting, Donald Fagan, and Wynton Marsalis, among many others. He mastered two Grammy-winning albums of the year, John Mayer's "Room for Squares" and Steely Dan's "Two Against Nature."

In this contribution, Scott Hull describes his view of full-range monitoring. One might notice that in every mastering suite of Masterdisk, high-quality full-range monitoring is used. Scott explains why he believes this to be a vital part of professional audio mastering.

The Case for Full-Range Monitoring

Scott Hull, Senior Mastering Engineer
Owner of Masterdisk Mastering Studios, New York, NY

Music is taken into the body and interpreted by all our senses. While our ears get all the credit, I'll bet that most of you have observed that music feels different in different environments. The vibration of the floor, the air pressure or thud of the bass in your chest, the spatial information that your eyes collect, the phantom images created by your brain, and the simultaneous reaction of others influence how you react to music. These opinions are based on what environment you are in, not only on what you hear. Many scientists have created elaborate trials to learn about human hearing, but one thing that I contend is true from my experience is that you cannot completely isolate your ears or your perception from other environmental factors.

One obvious alternative for monitoring music is to place the transducers right next to your eardrums so that you only hear the electrically produced sound and nothing from your environment. That may be valid, but only if you began listening to music this way from birth and never heard music played in a room where you experienced the complete picture. When we listen to headphones, our brain is trying to connect what it hears with what we have heard before. This pattern-matching phenomenon is well documented in our other senses: vision, taste, smell, touch. For instance, if something tastes like something we remember, then that *is* what it is. Many other observations are similar.

I feel that to truly react in an emotional way, you need to experience the overall impact from the music. Small speakers or even midfield-sized monitors can give you a very technical observation. But the impact of the entire spectrum, I think, is everything. For me, this emotional reaction is not as profound when played through smaller speakers.

Here are a few technical reasons why small monitors are less than ideal. Subbass frequencies below 50 Hz are very difficult to reproduce from a near field. To make these speakers seem like they have lower fundamental frequency response, the manufactures put ports and passive radiators on the speakers. Some of these work great at producing a very pleasant-sounding experience, but almost all of them do this at the expense of flat frequency response. The specs on these speakers claim flat response, but I find that only tells part of the story.

When mastering, we are constantly testing different settings. A-B-ing. If I think a piece of music could be improved with a little more bottom end, I will experiment with a decibel or so of EQ at several different frequencies. Often my instinct (guess) is right-on; other times it's a nearby frequency. But what I find very interesting is that on small speakers 1 dB of bottom EQ—for instance, around 70 Hz—might be completely inaudible. In other words, if I boost 1 db at 70 Hz with an equalizer, I cannot perceive the change—it sounds the same. And this usually leads me to over-EQing—adding more EQ to achieve the desired result. But when I'm curious and hunt around, I might find that on these speakers 1 dB at 62 Hz is completely audible. The dilemma is which is right? My presumption is that when using a ported enclosure to produce more dramatic low-end response, the resonant frequency of the port plays a role in the listening experience. While the speaker sounds like it has enough low-frequency energy in general, it may not be an accurate reference. You can get used to the sound, but if you cannot discriminate the difference, then it doesn't make a good reference monitor. The relative success of small speakers depends on the program material. In fact, the key of the song has a bearing on this. If the fundamental frequency of the tonic (root) pitch in the bass lines up with the port resonance, then it may appear to have way too much bass, causing you to reduce the bass energy in your music—but only on those notes (frequencies) closely related to the fundamental frequency.

What Is That?

Nearly every single day in my mastering room I am working on a record that I haven't heard ever before. It's a new experience each day. I love that about mastering. But one thing that is consistent is when producers and mix engineers attend my mastering sessions, they almost always say to me, "What is *that*?" They have just heard something new in their mixes that they never heard before. This isn't in itself a problem, but the reason they bring this up is because it's usually some sort of subtle defect that doesn't sound subtle at all in my room. This magnifying-glass approach is essential in mastering, and not only in the high frequencies. Without that clarity and full-range response, along with amplifiers that have a nearly infinite slew rate and very low noise floor, those subtle defects are masked. Even the little fan that keeps your computer cool produces enough background noise to make finding small clicks in the music almost impossible.

But What About the Consumer?

We all are very aware that the music consumer will listen on ear buds and laptop speakers under the keyboard and all sorts of other environments that are very far from ideal. So why go to such lengths to create an ultraquiet listening space, have high-resolution converters, enormous amplifiers, and speakers that are actually as tall as I am? The answer is simple: I need to be able to hear and feel everything, and this is the environment *I* am intimately familiar with. These speakers are *my* ears.

What Are My Goals in Monitoring?

I must be able to discriminate EQ and level changes down to the tenth of a decibel. For real—that is what I do on a daily basis. If your textbook tells you that you can only hear a change of ±3 dB, you are reading from a completely outdated book. When making changes in EQ in these tiny increments, the observation is almost always communicated as a feeling or an impression. "It feels like there is a bit more edge or air or presence or body or warmth." But collectively these supersubtle changes are what give my mastering dimension and emotion beyond basic sound quality and dynamics. If you cannot discern these minute differences, what are you really hearing? This is heavy stuff. You only get to this place by making mastering your life's work. Mix engineers and producers move from room to room and do a tremendously great job of adjusting to their environment. Mastering engineers, however, try to control our environment and use a singular monitoring approach. This way I know what I am hearing, and I can adjust for it immediately. It's been said that a (wo)man with two watches never really knows what time it is. I have to say that I feel exactly that same way about my monitoring system. One set of ears, one set of speakers—and amps—and cables—and one room.

Music is emotion.

Jaakko Viitalähde is a mastering engineer, electronics enthusiast, and founder of Virtalähde Mastering in Kuhmoinen, Finland. He personally custom-built and designed most of his studio facility and equipment down to the monitors, equalizers, and console. His remarkable experience and range of skill afford him a deep understanding of mastering processing and equipment.

In this section Jaakko Viitalähde explains his well-developed views about optimizing an analog signal chain for mastering.

Connection and Calibration of an Analog Mastering Chain

Jaakko Viitalähde, Owner and Mastering Engineer
Virtalähde Mastering, Kuhmoinen, Finland

Prologue

A mastering room installation seems like a relatively straightforward job on first look. Usually, there are only a few well-selected tools to connect together, and the required flexibility in routing is much less compared with tracking or mixing rooms. Yet, despite the apparent simplicity of the installation, great care must be taken in bringing it all together. The aim of any mastering room is to be capable of doing short, clean signal transfers and to bring out the best in the source material in the most intuitive way possible for the engineer. I will provide some food for thought and discuss ways to improve workflow.

Building Up a Chain

A mastering chain is never just a random collection of tools thrown together. There is a reason they are built like they are, and it all comes down to the engineer's matured personal preferences and ways of working. While maximum flexibility and freedom in patching might at first seem like the perfect idea, there is a very good chance that patterns will be found in workflow after the equipment has been around long enough and a fair amount of work has been done with it. By being aware of your own working methods and by actively challenging yourself, any mastering chain can be effectively streamlined.

The minimum requirement for any piece of equipment used in mastering is to sound good when set flat. Sound, or tone, is a highly subjective topic, but for a sonically transparent tool, the tone of a unit with no actual processing applied should be the same as the source or better with any source material. If the tone is "almost there," it makes no sense to keep such tools in the chain unless used only occasionally and there is no substitute for what the actual processing does. Tone tends to cumulate and eventually may lead to compensating for the slightly inferior tone of the chain with unnecessary processing.

Having tools with a tone that works with only some incoming material is perfectly fine. The core of the mastering chain should, however, work with just about anything that needs processing. Transparency goes a long way with a core chain that is both sonically transparent and a little bit better sounding with any incoming material.

A/B comparisons performed with the digital audio workstation (DAW) are an effective method for evaluating tone in an objective way. Bypassing a unit can tell you something, but judgment is still affected by small changes in level and even by differences caused from interaction between units. Some outputs sound different when driving various inputs, so a unit bypassed in between other units does not necessarily provide a clear perception of how the unit sounds at a specific spot in the chain. Properly implemented, A/B comparisons remove every such variable and provide the freedom for trying out different kinds of combinations without being misled.

By capturing the sound of the chain and its various parts or combinations back into the DAW, the carefully level-matched files can be blindly compared to the source material. The results can be brutal, and there is a good chance of finding out that the sound of the original files is preferred. It is true that applying actual processing changes things, but even with the desired processing used, the same basic tone will always be there. If the source material needs some EQ, is it better to have a sound that is "almost there" with the additional EQ or one that sounds the same or better plus the applied EQ? At the last critical stage of mastering, every little detail counts.

Good-sounding individual tools are only the beginning, and it takes a lot of time and effort to build a balanced-sounding mastering chain. A large part of this is finding tools that work well together, making the combination greater than the sum of its parts alone. As discussed previously, the core of any mastering chain should be fairly simple, transparent, and flexible enough for most of the daily workload, and it is the perfect place to start. When you have a chain that sounds good on just about everything, adding tools with a tone that fits only some material is much easier.

The overall patching order is something that evolves over time. While some consoles do allow for total freedom in patching, parts of the chain often tend to end up in a fixed order because they simply sound better that way. This depends on how the inputs and outputs are constructed, and the only way to find the best arrangement is to carefully listen to how various configurations sound. Because the patching order is affected both by personal preferences and by the fact that one combination sounds better than the other, experimentation is required for finding a good balance.

Patching Methods

The simplest way for patching is daisy-chaining (also called *point-to-point wiring*). Patching a chain from one unit to another is a common method and can be an extremely pure way for building up the signal path. Such a method works very well when the number of tools is fairly limited and the patch order rarely or never changes. The interconnections can be kept very short, and the number of mechanical contacts on the signal path is kept to a minimum. Obviously, as the amount of equipment grows, the overall chain gets rather complex to handle, and the signal travels through a lot of contacts, even when bypassed.

A fixed patch relies entirely on the quality of the individual local bypasses. Not all of them may be true bypasses, which by standard removes the circuit completely from the signal path. This is the electrical equivalent of connecting the XLRs back to back. There are also bypasses that may keep the input active at all times and just switch between the input and output, keeping an extra load on the preceding amplification stage at all times. Another possible bypass is that which keeps some circuitry in the signal path, such as the I/O amplifiers. Knowing how the bypasses are built is essential when building a point-to-point patched mastering chain.

For more flexibility in routing, mastering consoles are often used. They come in all shapes and sizes, passive and active, and while the small details may differ quite radically, at their core, most of them offer some type of inserting system. The number of available inserts may vary from three to eight or so. The less available inserts there are, the more likely the user will be daisy-chaining several pieces of equipment together inside an insert, and with more inserts, each insert will serve as a clean, short way for bypassing each tool individually. Most consoles have passive inserts, which should, in theory, be as transparent as possible. This also makes the installation much more dependent on how the individual tools interact with each other. If the inserts are actively buffered, this dependency is broken, but buffering does add a few extra stages of potentially unnecessary amplification to the chain.

With a wider selection of tools at use, a good and flexible mastering console is an elegant way for controlling it all. Many consoles offer additional features such as M/S matrixes and ways for controlling the patch order within the console. Bypasses are clean, with each located in the same place, and the signal does not have to travel through every piece in the desk if the required job at hand needs, for instance, only one EQ. However, the opposite is true for more complex patches because there is increasingly more back-and-forth cabling and extra contact points in the signal path compared with a similar fixed patch.

Patchbays are another option, but they are rarely used in mastering rooms. While a patchbay does give good flexibility in routing and can be just as good as any other patching method when properly installed, patchbays are not often necessary because mastering chains tend to be quite simple and fixed in order, even when wrapped around a mastering console.

The interconnections used should always be built from cables and connectors of good quality and be as short as possible. By clever physical placement of the equipment in the racks, the length of the interconnections can be easily customized. Arranging your tools cleverly also has an effect on your workflow. Easy access to everything and a sense of balance in the racks promote a good working mood.

Installation, Calibration, and Operating Levels

A typical mastering chain might be a mixture of both balanced and unbalanced interconnections. Some commonly used tools in mastering come with unbalanced inputs and outputs, but because the cable runs are usually short and induced noise is not typically a problem in mastering rooms, there is nothing wrong in using unbalanced interconnections. In fact, running unbalanced can in some cases sound better if the inputs and/or outputs are modified and balancing/debalancing stages are bypassed or removed, transformers in particular. Such modifications are another step in tuning a mastering chain for its best possible performance.

Inputs and outputs can be balanced both electronically and passively by using transformers. Transformers are excellent for isolation and trouble-free operation, and they can be used either balanced or unbalanced, but they are never completely transparent in tone. Transformers can be a little picky on the preceding line driver or the load after them, and there will always be low-frequency distortion and overall phase shift present to some degree. Some transformers do sound excellent, but not all of them. When using a transformer-balanced input or output in an unbalanced system, the cold pin (usually pin 3) always has to be grounded or there will be a noticeable loss in level, especially at low frequencies. If the connection to the transformer does not have its pin 3 grounded, it has to be wired to the ground (pin 1) in the interconnection cable or by modifying the I/O stage.

Electronically balanced inputs and outputs are usually quite easy to mix with unbalanced circuits. As always, pin 3 must be grounded when operating a balanced input unbalanced, or the unconnected input at pin 3 might induce noise in the circuit. Some balanced (nonfloating) outputs do not like their pin 3 grounded, and there can be 6 dB of loss in level or even damage to the output circuit. If such an output has to be connected to an unbalanced input, pin 3 needs to be left unconnected at the other end. Another solution would be to use an extra debalancing stage, which would also solve the level-drop problem.

A well-implemented grounding scheme can reduce low-level hum and buzz in any installation, and mastering rooms are no exception. A completely balanced chain is usually very tolerant of slight imperfections in grounding, but unbalanced connections could require extra care. With some equipment, problems can emerge if pin 1 is grounded directly to the circuit ground instead of the chassis ground, as described in Rane Corporation's note, "Grounding and Shielding Audio Devices," by Steve Macatee (www.rane.com/note151.html). This may result in hum and buzz problems. Without modifications, the only thing to try is to disconnect pin 1 from the interconnecting cable at the signal-grounded end. The referenced Rane note is an excellent resource for any installation and its potential grounding problems and is worth reading by anyone.

Usually, impedance Z is not a thing to think of when connecting modern studio equipment together. The outputs are low Z and the inputs are high Z, so the output drivers have no trouble driving a high-Z input. Lowest nominal input impedances tend to be in the 2.5-kΩ ballpark, which should be plenty for most outputs to drive.

However, some transformer-balanced inputs may represent a more complex load that demands more from the preceding output stage to sound the best. This explains in part the differences heard in various combinations of equipment, and the only way to find what sounds best is to test all possible combinations that make any sense at all.

Pre-1960s equipment is relatively rarely used in mastering, but when such equipment is connected to a modern line-level system, it is mandatory to thoroughly check whether it is designed to work in a 600-Ω installation. This means that the inputs are of low impedance and need a strong driver stage before them and also that the output needs to be terminated properly with a 600-Ω resistor if the output connects to a modern high-Z input.

One of the most important and often overlooked aspects is the operating level of the system. Any mastering chain needs to be calibrated to a chosen reference level. There are no standards to follow, but in general, low levels work well with analog equipment, and when there is enough headroom available throughout the chain, the sound will be much cleaner. There is a time and place for driving parts of the chain with higher levels for various saturation effects, but for the most part, using operating levels that are near the clipping point only does harm to the audio.

For example, an often seen reference for 0 VU (or +4 dBu) is −18 dBFS. This leaves us with only a tiny reserve of headroom because 0 dBFS sits at +22 dBu, which comes close to the clipping point of some of the generally used equipment. By choosing a dramatically lower operating level, such as −8 dBFS, +4 dBu, 0 dBFS is +12 dBu, leaving us with 10 dB of available headroom for equipment with a clipping point at the mentioned +22.5 dBu. Many line-level amplifier circuits are grossly nonlinear near the clipping point, but it only takes a few decibels of safety margin to considerably clean up a chain. Naturally, by driving the chain less hot, you are also operating closer to the noise floor, but since mastering equipment should, by standard, be of low noise, this is usually not a problem.

Closing Words

A large part of getting into the flow is having a set of tools that do not require much thinking while working. Spending a long time on one track tends to blur the judgment, which is why it is crucial to thoroughly know how your mastering chain works in its various configurations. When everything has found its place, there is no second-guessing, and the important decisions can be made quickly.

The tools of a mastering engineer truly are a balancing act. It takes time and patience to build a mastering chain, and it is as delicate as mastering work itself.

Dave Hill founded the highly respected equipment manufacturing companies Crane Song, Ltd., and Dave Hill Designs. He is one of the foremost equipment designers in the world of recording and mastering and is the subject of the documentary film *Crane Song Superior Gear: The World of Dave Hill*. His designs can be found in most modern professional mastering studios as well as many recording studios across the globe. He is especially known for the quality of his harmonic distortion processors, which are so renown that a version is now included in ProTools.

Dave Hill helps us to understand how harmonic distortion processing can be used to shape tonality.

Distortions and Coloring

Dave Hill, Audio Engineer and Owner of Dave Hill Designs
Equipment Designer for Crane Song, Ltd.

Introduction

Distortions come in many forms, ranging from obvious to subtle. Some are useful, whereas many simply sound bad. When distortion is used for color, it can be a powerful tool, but it must be used carefully. In precise terminology, anything that alters a sound could be called *distortion*. This includes such common tools as EQ and compression. Among other uses, EQ is often thought of as a correction tool, allowing us to correct for distortion and inaccuracy. Compression/limiting, when used to make things loud, creates a distortion that is not pleasing to the ear. Limiting was first designed to control infrequent peaks. When used as something to make the source louder, with almost constant limiting, it is not a good thing. The ear thinks louder is better at first listen, but a more realistic perception about this can be had by listening with carefully matched loudness levels or from extended listening. We begin to hear the unpleasant distortion this way.

Think about human hearing. Nature does not have continuous, loud sound sources. Hearing most likely developed for communication/speech and for survival. If all that is heard is high-energy sound, for which the ear did not evolve, what are the long-term consequences?

There are many cases where restricted dynamic range is needed so that the source can be heard, but the resulting sound is not ideal. There are many records that have done very well without having the life compressed and limited out of them. An example is "Dark Side of the Moon." It has dynamics, something that is being lost in modern recordings. The trend has been to make things play louder and louder, and the result is deteriorating sound quality. Such undesirable distortions should be avoided.

Distortions as Color

Hearing is quite an odd phenomenon and is not very well understood. It turns out that distortions may not be heard as distortion. For example, when adding second harmonic distortion to a source, because it is an octave musically, it can be very hard to hear. However, it does change the character of the sound. Some people cannot hear a difference even when large amounts of second harmonic distortion are added. On a pure-tone instrument it may give a bad sound, but on a complex sound there may be an enhancement. "Euphonic" is how some high-end hi-fi enthusiasts describe the sound of their tube amplifiers that produce a significant level of second harmonic distortion.

There are many examples of color with analog gear. People may like a certain piece because it has a warm sound. Is this sound due to a nonflat response—a high-frequency (HF) or a low-frequency (LF) boost? In many cases it is frequency-response errors that we hear first, not harmonic distortion. Some people describe a certain D/A converter as having a warm sound, whereas the sound is actually the result of distortion introduced by filter problems.

Whereas the character of a piece of audio equipment could be due to a frequency-response error, harmonic distortion also could be responsible. For example, third harmonic distortion and other odd-order harmonics can produce a bright sound. It can be a desirable sound with the right input level and with lower-order harmonics. On the other hand, higher-order odd harmonics tend to be harsh-sounding. Generally, it is harmonics above the fifth that are often referred to as *higher order*, with the lower ones being called *lower-order harmonics*. When one peak limits and turns a source into a square waves, the distortion we have is in the range of high-order odd harmonic distortion, which is bright and harsh.

Magnetic tape primarily introduces third harmonic distortion, but there are distortion-level versus frequency-response considerations and system frequency-response errors—head bump, HF roll-off, and many other effects take place. On tape, the distortion level will increase with frequency owing to reduced HF headroom, although, because of the response of the playback amplifier and other factors, we do not hear the HF part of the distortion very much. Also, because usually the HF content of a recording is not large, what we hear is mostly the LF and low-midrange distortions, the time-domain errors, and the effect of low HF headroom on transients. The point is that just because we can introduce harmonic distortion as a coloring effect, it is not always good or predictable. Part of the charm of analog gear is the unpredictability, unlike the digital domain. The part of tape we like is the warming of the midrange with harmonic distortion. There is some transient reduction, but if you push tape to seriously remove transients, the high end becomes dull. Overall, a good tape machine used with typical levels will not change the sound very much.

Nonlinear

With analog gear, the mechanism that creates harmonic distortion is an error of linearity—an error of the input-to-output relationship. This nonlinear relationship varies based on the signal level and frequency. Take the idea of soft clipping—the tops and bottoms of the signal are being rounded off instead of chopped off. Because this is taking place on both the top and bottom halves of the wave shape, we are adding odd harmonic distortion. The shape of the rounding will affect the level and type of harmonics that are generated. The function taking place is symmetrical, meaning that the top and bottom of the wave shape are affected in the same way.

Analog gear will have something between a very hard clip and a soft clip depending on the circuit type. Integrated-circuit (IC) amplifiers tend toward the hard clip, whereas tube amplifiers tend toward the soft clip. One must be aware that this is very general and that each amplifier has its own characteristics.

Now if we round off only the top or bottom of a wave shape, we have asymmetrical distortion, and it will be mostly even harmonic distortion.

The shape of the nonlinearity changes the distortion. The generated distortion may or may not be heard as distortion or even heard at all. If you add frequency-response-dependent nonlinearity to this, you may have a piece of gear that could be very clean until 5 kHz and above or perhaps clean above 100 Hz.

The generated distortions on the low end could sound like a fat, warm sound, or it may make the low end more undefined sounding, mushy, although with some types of music this may be a good thing. Distortions above 5 kHz may sound like "air" or brightening, or instead it could sound edgy and bad. Different source material with various levels will be affected differently. What happens in assorted analog gear can vary widely. No two units, even of the same type, will have exactly the same character. There will be small variations.

Along with the generated harmonic distortions, there is intermodulation distortion.

This is mostly a bad thing. It is the generation of non-musically related tones into the program. If the nonlinearities are soft enough, the amount of intermodulation distortion may be so low that it is not audible.

Time Domain

Distortions in time are a bigger problem than is recognized. We often think of transients and wanting to smooth them or slow them down, and doing this can make some material more listenable. It is also thought of as something we must lessen so that the music can be louder. This takes life, "air," and space out of music if pushed to the limit. This is perhaps the most well-known and widespread problem in the time domain.

Time-domain problems that are not so apparent to those who have never heard anything other than 44.1k recordings relates to imaging. This is known as *pre-* and *postecho.*

Pre- and postechos of the program source are added to the program when working with digital audio. It is most apparent on transients, such as guitar strums. This effect is known as the *Gibbs phenomenon* and is part of the physics of digital audio. The linear phase filters used in converter circuits create the pre- and postechos, disturbing the imaging. There are several techniques circuit designers can employ, such as the use of minimum-phase filters, but it is impossible to completely avoid the phenomenon. These filters, used for antialiasing, are a required part of both A/D and D/A conversion and are a major source of time-domain issues. Given the steep filter requirement of the 44.1k sample rate, it is not possible to avoid linear-phase filters without undesired side effects. The only real way to improve this is to use higher sample rates; 192k sources image much better than 44.1k sources. It turns out that processing at 192k also sounds better. There are many benefits to working at the higher sample rates, even if the end will be 44.1k or worse. If the source sounds better to start with, then the end result will be better. Processing with analog gear or other effects will not remove the time-domain errors once they are present. For this reason, the number of conversions is often kept to a minimum.

Coloring

When dealing with a recording that needs work to make it more presentable, there are the standard tools (i.e., EQ, compression, reverb effects, M/S techniques) as well as tools that change the harmonic content. These harmonic tools work by introducing analog-like errors to the digital source. If used skillfully, they can affect the sound in a positive way, making it sound less "digital." The better processes will be a bit unpredictable because the result depends on the level and frequency content of source material, just like analog gear. One can use a tape-like emulation to fatten the lower midrange, which is not possible with an EQ. If ones adds third harmonic distortion of the correct type, the sound can become brighter and bring out detail.

If a compressor limiter is used to raise the level of a sound, it is usually best to perform the coloring first, before the dynamic range reduction. If coloring is done with something other than EQ on a source that has little dynamics left, it must be done carefully to prevent displeasing distortion.

The best results with harmonic processing are with open material with space to allow the altered harmonic structure to be heard.

Brad Blackwood is the owner of Euphonic Masters in Memphis, TN. His remarkable mastering career began at Ardent Studios in Memphis, TN, where he revived the Ardent Mastering brand. He has over 3,000 album credits, including work for artists such as Maroon 5, Black Eyed Peas, Everclear, POD, North Mississippi Allstars, and Three Days Grace. His work has received several Grammy nominations. In 2012, Blackwood won a Grammy for Best Engineered Album for his work on Alison Krauss and Union Station's "Paper Airplane."

Brad Blackwood shares with us his views and techniques for mid-side processing. This provides a glimpse into the ideas developed by a professional who has successfully integrated this processing into his everyday approach.

Mid-Side Processing

Brad Blackwood, Mastering Engineer and Owner
Euphonic Masters, Memphis, TN

While a vast majority of my work is done via standard stereo processing, I do use mid-side equalization quite a bit (EQ is the only processing I use in M/S mode because I've not heard mid-side compression that sounded right to me). I do this in the analog domain, using a custom-built analog M/S matrix along with the Crane Song Ibis EQ, which is perfectly suited for mid-side work because of its overall flexibility.

Most balancing and frequency-response issues are best addressed via stereo EQ, but there are times when I feel that mid-side EQ allows me to more transparently tweak the audio. For example, if you have a nice vocal/snare sound but the hard-panned guitars are dull, mid-side EQ can allow you to open up the guitars without affecting the vocal and snare. Likewise, if you have a mix where the vocal or snare needs to come up a touch, those elements can be enhanced via middle-channel EQ without applying unnecessary equalization to the entire mix. M/S processing also can be used to enhance a mix's perceived width by emphasizing hard-panned details such as reverb.

I most commonly use M/S EQ to enhance guitars and other panned instruments, although I sometimes remove excess hard-panned bass frequencies in order to tighten up the bottom end. It's also not uncommon for me to open up or darken the vocals a bit in the middle channel—an area in guitars and synths that often doesn't need the same tweaks. Besides the flexibility the Crane Song Ibis gives (with its ability to do both broadband and very tight, "surgical" tweaks), I find the ability to increase the overall density of the middle and side independently via the color modes very useful.

While I consider mid-side EQ to be an indispensable tool, as with anything else, there are limits to what can be achieved—it's not some sort of magic bullet. Every bit of processing we do in mastering has both positive and negative impact on the audio; our job is to ensure that we make the choices that offer the greatest positive while minimizing the negative.

Pieter Stenekes is an audio engineer, software developer, and founder of Sonoris Software in Damwâld, Netherlands. Sonoris Software is one of the most respected audio plug-in and mastering-related software companies. This includes the highly regarded Sonoris Mastering Equalizer and Parallel Equalizer plug-ins.

Equalization is perhaps the most important processing in audio mastering. Pieter Stenekes helps us to understand ideas behind the equalization filters that can affect how they are applied by mastering engineers.

Digital Filtering

Pieter Stenekes, Founder and Owner
Sonoris Audio Engineering, Friesland, Netherlands

What Is a Filter?

Filtering is a broad topic. In general, a filter removes something from whatever passes through it. In this article I will limit the subject to digital filters such as those used in equalizers. These types of filters remove unwanted frequency components from a digital signal to enhance a piece of music. There are many types of digital filters, and each has its strengths and weaknesses, especially when it comes to audio, as will follow shortly.

Digital Signal Processing

Before we can talk about digital filters, we have to know a little bit about digital signal processing (DSP) in general.

Analog-to-Digital Converter
The single most important aspect of the digital domain is that a digital signal is not continuous like analog but discrete. To transfer an analog signal to the digital domain, it has to be sampled with an analog-to-digital (A/D) converter. This

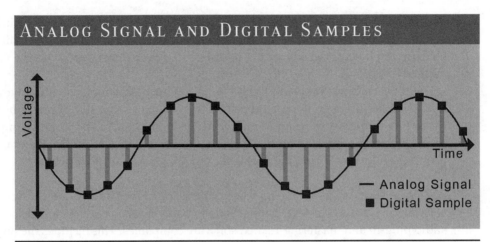

FIGURE 16-1 Analog signal and digital samples.

device takes in a sample at discrete time intervals (*discretization*) and replaces it with a sample value (*quantization*). This process is illustrated in Figure 16-1. The frequency at which this happens is called the *sampling frequency* or *sample rate*. Popular sample rates in audio are 44.1 or 48 kHz—but how high does the sample rate have to be? This is defined by the so-called Nyquist-Shannon sampling theorem. This theorem states that a signal can be perfectly reconstructed if the sample rate is greater than twice the bandwidth of the signal. Now you can see why the popular audio sample rates are higher than approximately 40 kHz—because the practical highest frequency a human ear can hear is 20 kHz! The other way around is also true: A signal to be sampled may not contain frequencies above Nyquist; otherwise, weird signal components will be generated, known as *aliasing*. The signal must be filtered first.

Digital-to-Analog Converter

Once the signal is in the digital domain, it is just a sequence of values that can be stored and transferred easily and without any quality loss, unlike an analog signal. And, of course, the signal can be manipulated in various ways. But sooner or later the signal has to be converted back to the analog domain with a device called obviously a digital-to-analog (D/A) converter. A D/A converter raises the output voltage to the value of each sample until the next samples hold. This produces a staircase-like signal, as shown in Figure 16-2.

Although this signal looks a lot like the original signal, the hooked shape creates frequency components (harmonics) above the Nyquist frequency. Of course, this is not what we want, and therefore, we need to filter all frequencies above the Nyquist frequency to perfectly restore the original signal.

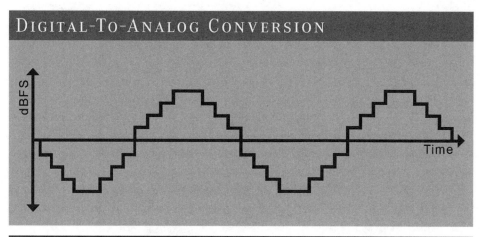

FIGURE 16-2 Digital-to-analog conversion.

Impulse Response

An impulse response is the response of a system to a very short signal, called an *impulse*. Why is this important, you might ask? Well, with this impulse response you can simulate the reaction of this system to a given signal. This process is called *convolution*. A popular implementation is a convolution reverb. Loading the impulse response of a reverb device into a convolution processor together with a dry audio signal will result in a reverbed audio signal, just as if the signal went through the real thing. To get a perfect simulation, the system has to be linear and time-invariant. Needless to say, the greatest sounding reverbs are far from linear and time-invariant.

Filter Types

IIR

This stands for *infinite impulse response*, infinite referring to a never-ending feedback loop. This filter type is very similar to an analog filter; in fact, you can quite easily transform an analog filter into an IIR filter with a mathematical process called *bilinear transformation*. Figure 16-3 presents a diagram of an IIR filter. If you feed an IIR filter with an impulse signal, it will output an infinite number of nonzero values.

The output Y is fed back through H to the input. This inherently means that anything happening on the input will theoretically be present in the output forever, hence the term *infinite*. In practice, the response of a stable system will fade away into the noise.

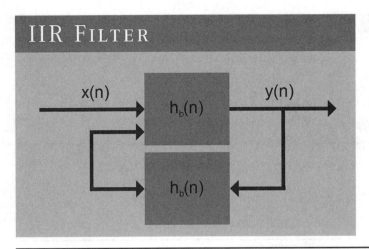

IIR FILTER

x(n)   y(n)

$h_b(n)$

$h_b(n)$

FIGURE 16-3   IIR filter.

One of the problems that can arise with an IIR filter is instability because of the feedback. Mathematically, this can be explained as shown in Figure 16-4.

H is the transfer function. For a very simple IIR filter, H may look like that shown in Figure 16-5.

A problem arises when the frequency component Z results in a division by zero! In this simple filter we have a 1 pole at $Z \times A$. With careful placement of the poles, such a filter can be made stable.

IIR FILTER: RELATIONSHIP BETWEEN X AND Y

$$Y(z) = H(z) X(z)$$

FIGURE 16-4   IIR filter. This figure shows the relation between input X and output Y.

IIR FILTER: TRANSFER FUNCTION

$$H(z) = \frac{1}{1 - a/z}$$

FIGURE 16-5   IIR filter transfer function. Z is a so-called complex number and is another way to describe the frequency. A is a property for this given IIR filter.

As mentioned earlier, an analog filter can easily be transformed into an IIR filter. As a result, most classic analog filter designs have a digital counterpart. Each filter has its strengths and weaknesses. Here are a few popular design types with their typical characteristics.

Butterworth Filter   This is a so-called maximally flat filter design with a smooth transfer from the pass band to the stop band. The frequency response in the pass is flat.

Chebychev Type I Filter   This has a steeper transition between the pass band and stop band than the Butterworth filter. The stop band is as flat as possible, but the pass band has a ripple.

Chebychev Type II Filter   Again, this has a steeper transition than the Butterworth filter. In contrast to the Chebychev type I filter, this type has a flat pass band and a ripple in the stop band.

Elliptical Filter   This has a very steep transition between the pass and stop bands but with a ripple in both bands.

Most IIR filters start behaving differently than their analog counterparts around the Nyquist frequency owing to a mathematical effect called *prewarping*—but what if you also want an accurate response at these frequencies? A common approach is the use of upsampling. Here the audio gets upsampled to a higher sample rate, moving the problem far above the original Nyquist frequency. After that, the signal is downsampled again to the original sample rate, and the problem is gone. Of course, implementation of the upsampling algorithm needs to be very precise; otherwise, the end result is worse than before.

FIR

A FIR filter has a *finite impulse response*. This means that the response Y to an impulse X will last a predetermined time and after that becomes zero. As you can see in Figure 16-6, a FIR filter doesn't have a feedback loop like an IIR filter. This means that a FIR filter is inherently stable.

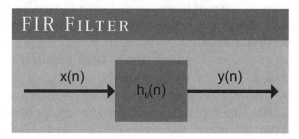

FIGURE 16-6   FIR filter.

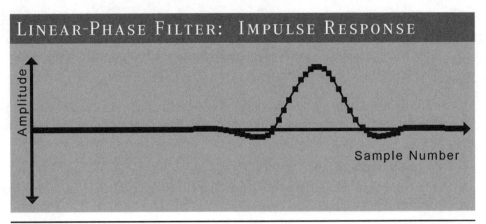

LINEAR-PHASE FILTER: IMPULSE RESPONSE

Amplitude

Sample Number

FIGURE 16-7 Linear-phase filter.

FIR filters are relatively easy to design and implement, especially in hardware. Another advantage of a FIR filter is that it can have exactly linear phase. A system has linear phase when its impulse response is symmetrical. Figure 16-7 shows a linear-phase filter.

A FIR filter can have this property very easily because the filter impulse response is defined directly in the design. With an analog or IIR filter, this is impossible because of the feedback loop causing the impulse response to depend on what happened in the past. This is why you won't find linear phase in nature. Also, with a linear-phase filter it takes a while before the input signal reaches the output, and this causes a delay. Some people find this reason enough to disqualify linear-phase filters for audio, whereas others find these filters to be very transparent because there is no phase distortion.

Are there any other disadvantages to FIR filters? Of course there are. For example, a FIR filter needs a higher filter order than an IIR filter to get the same response. This means that it needs more memory and computational effort.

Design and Implementation

In equalizers you can find various types of filters. Low- and high-pass, peaking, and shelving filters, and so on, whether FIR or IIR . While it is hard to generalize, I think it is safe to say that a digital filter needs to be stable under all conditions and implemented properly before it can be used successfully as an equalizer or any other filter.

Stability is relatively easy to achieve when carefully designed, but proper implementation is harder to do. This depends largely on the chosen hardware platform, floating- or fixed-point arithmetic, engineering skills, and so on. This is where manufacturers can really make a difference. A filter design may look fantastic on paper, but when implemented poorly, it will fail to produce excellent results.

In an analog equalizer, the electronic parts are very important. There are many types, and there is a lot of difference in quality between brands. Also, capacitors will change over time. In the digital domain, we don't have this trouble, but unfortunately, we have to deal with other problems. We have to take care of resolution, sample rate, jitter, aliasing, and so on.

Conclusion

We have just scratched the surface of digital signal processing here. But there are many more possibilities, and we have to embrace them. When it comes to filters, in the digital domain we can simulate analog filters, create new ones, and also make filters that can't even exist in the analog domain. With proper design and implementation, a digital filter can work as good as an analog filter and even better!

Cornelius Gould is a former broadcast engineer most recently for CBS Radio in Cleveland, OH. He now is the team leader for the Omnia audio processor brand and is the codeveloper of the Omnia 11 broadcast processor, used in most FM radio stations across the United States.

We learn from Cornelius Gould about how we can best prepare recordings for radio play and avoid known problems.

Optimizing Audio for Radio

Cornelius Gould, Audio Processing Developer
Omnia Audio, Cleveland, OH

There are several things that take place when a recording is played on the radio. First, general managers and program directors typically make a decision about which recordings a radio station will play. Then the station will download the music into its systems from a broadcast music download service (such as Play MPE and New Music Server). Then, for tracking and royalty-payment purposes, the station reads the ISRCs into its system, which may come from a database or be entered manually. ISRCs are virtually never extracted from any metadata or disc. The station also may obtain a CD of the recording, but typically music arrives by means of a digital delivery service. In the future, broadcast music download services may begin accepting and distributing 24-bit masters for radio, especially now that such masters are already being generated by mastering engineers for Apple's Mastered for iTunes program. I know of some broadcasters who already use 24-bit recordings exclusively, although they personally work to obtain them.

Today broadcasters concerned with maintaining quality source material mostly use 16-bit/44.1-kHz WAV files for broadcast.

Radio stations have their own unique selection of processors in front of the broadcast transmitter. There are many such processors typically called a *broadcast processors*, such as our line of Omnia radio processors.

Before the recording is sent to the transmitter, it must be processed in accordance with the legally instituted preemphasis curve—an equalization filter. The *preemphasis* curve is an early form of noise reduction still in use today, in which the high frequencies of an audio signal are boosted by some amount for transmission over a delivery system. The reverse of the curve is applied inside the listener's radio, which is called *deemphasis*.

This process creates brief high-frequency "peaks" that are applied to aggressive peak-limiting processes to reduce the peak energy. This process historically has been performed using clipping processes, although some processors employ some degree of sophisticated dynamic limiting action as well.

When a mastered recording is overly bright, the preemphasis processing can have a negative effect on the perceived on-air quality of the recordings. The peak energy of the high-frequency energy can become excessively high, causing a "smeared" sound to the high-frequency components. In some cases, severe clipping artifacts on the sibilant sounds, cymbals, and other high-frequency content are heard. Overly bright recordings have become more prevalent, especially with the limitless nature of digital recording, and generally were at a better level for radio during the early 1990s.

Broadcast processors perform various processing functions, including automatic gain control (which mimics manual gain riding and operates over a large dynamic range), multiband leveling, dynamic limiting, equalization, preemphasis filtering, brick-wall limiting, and more. They are typically set up by the manufacturer or the radio station's engineers. This processing is a type of automatic/algorithmic mastering, if you will. Sometimes anomalies happen with this compression, as with Britney Spear's "I Wanna Go." This song contained a relatively loud 20-kHz tone in the chorus. This tone causes higher amounts of gain reduction on the high-frequency bands in many multiband broadcast audio processors. This tone unintentionally causes the chorus to sound dull (like AM radio) compared with the verse on the radio.

Other songs (such as Bruno Mars' "It Will Rain") have significant subsonic components (around 3 to 10 Hz). These components cause some radios to mute during the intro because the radios think they are not tuned to a frequency given that the subsonic energy looks similar to a condition that translate to the radio as meaning "I'm not tuned in."

A useful tool to have during mastering would be some sort of fast Fourier transform (FFT) audio spectrum display to ensure that there are no unintentional tones or noises that fall at the extreme frequency-response edges that may cause a problem for songs when presented to radio stations.

Phase rotators are very typically used in broadcast. These help to increase the symmetry of a waveform, especially vocals. When a waveform is recorded, especially a voice, it can create an asymmetrical waveform. When clipping/brick-wall limiting techniques are applied, the sound quality is harsher if the source is asymmetrical. Phase rotators always degrade the audio to some degree but provide a benefit of more symmetrical waves and provide a less abrasive, pleasant experience for the listener.

CDs that are mastered with hard clipping for greater loudness can sound extra distorted on the radio because the clipped waveforms in the recordings are "spun around" in the phase rotator and can cause overload distortion in the front end of some audio processors. It is generally better to use look-ahead limiting where possible versus hard clipping.

Radio broadcasters are engaged in something not unlike the loudness wars of the mastering world. Stations compete between themselves with loudness levels stemming from the same reasoning behind the loudness wars. This means that broadcast processors must raise loudness, which is done primarily by clipping. The automatic/algorithmic mastering contained in the broadcast processor uses some of the same types of tools used by mastering engineers. Because of this, if mastering engineers leave some dynamics for the broadcast processor to work with, it will result in a better experience for the radio listener. Perhaps in the future this may be done by mastering engineers delivering a different version to the broadcast music download services, but today everyone gets the same version. Songs mastered with a maximum average loudness level of –8 or –9 dB root-mean-squared (RMS) may be ideal for radio playback. This would be according to a true RMS meter such as a Dorrough loudness meter, which would be –11 or –12 dB RMS on most other digital meters such as the Waves PAZ and those built into most digital audio workstation (DAW) software, which uses the more common RMS + 3.

In conclusion, there are several things mastering engineers can do to ensure the best sound quality on the radio. Mastering engineers should be careful not to make recordings overly bright or sibilant, which could degrade the sound quality owing to the preemphasis curve. Because of the loudness competition between radio stations, dynamic recordings will simply sound best on the radio and will be as loud as overcompressed recordings. Finally, mastering engineers should consider providing less "processed" versions of songs, and maybe even higher-bit-rate source material should be used by radio stations in the future, which is entirely possible with today's systems.

Jeff Powell is a veteran recording/mix engineer, producer, and vinyl-cutting engineer. Powell has enjoyed the success of multiple gold and platinum records, including five Grammy Award–winning projects. His credits include albums such as Stevie Ray Vaughan and Double Trouble's "Live at Carnegie Hall," Big Star's "In Space" and the Afghan Whig's "Gentlemen." He first began to learn the skill of vinyl lacquer cutting from legendary vinyl-cutting engineer Larry Nix, who began as the house mastering engineer at Stax Records and later founded L. Nix Mastering. Powell now performs direct vinyl transfers on a classic Neumann VMS70 lathe.

Jeff gives us the scoop on how to prepare a recording to be cut to vinyl. He shares details about the process with a focus on ideas that could affect a mastering engineer's work.

Premastering for Vinyl Cutting

Jeff Powell, Direct Vinyl Transfer Engineer
Engineer for Stevie Ray Vaughn, Bob Dylan, and Many More Artists

There are a few key things to understand to best prepare recordings for master vinyl lacquer cutting. First, let's cover a general overview of the process from the final mix in the studio to putting a finished record on the turntable for a listen. Later there will be more detail about the various steps.

The band, the mix engineer, and the producer mix the record, assemble a final sequence, and send it to their mastering engineer of choice. The mastering engineer processes the mixes and then assembles a final *Red Book* CD master and/or digital master. The CD master is then sent to the label or pressing plant to be manufactured. If the recordings are also intended for vinyl release, the mastering engineer will create another set of files optimized for cutting vinyl that are used by the cutter to make a master lacquer. The master lacquer is cut, inspected, and immediately shipped to the plating and pressing plant (plating and pressing services are usually provided by the same company). After the plating process, metal parts are created and used to stamp out five to seven test pressings, which are sent to the band, label, original mastering engineer, management, and the cutting engineer. Everyone listens, and if everything sounds great, the records are pressed.

There are several suggestions mastering engineers can make to the mix engineer and producer to get the best results from vinyl. This includes several things to avoid: overcompression of the mix, hard-panned drums or bass, piling on the giant low-end subfrequencies, and excessive brightness in the high-end frequencies (especially watch the hi-hat and cymbals for harshness and de-ess the vocals). In the most general sense, if both the mixer and mastering engineer are aware of these basic things, it will provide a big head start in making a great master source for the cutter to work with.

There are also some things a mastering engineer can do to help create the best-sounding vinyl. The most important thing to realize is that mastering processing doesn't have to change much, except for the final stage. The mixes should be processed as normal for a CD master, more or less, but at the end, the digital limiter should not be used. The level should be reasonably near 0 VU. This can be a real challenge if clients are turning in mixes with no dynamic range.

Luckily, most of my projects come in at 24 bits/96 kHz. It is always best to provide the cutter with at least 24-bit depth recordings regardless of the sample rate, whenever possible. The worst thing, next to someone asking me to cut an MP3 file to vinyl, is to give me a final *Red Book* 16-bit/44.1-kHz finished CD to fit onto a vinyl record the best I can. When this happens, the first thing I have to do is turn the level down by 7 to 10 dB to even come close to making it fit. There are physical limits to what can fit onto a side of a vinyl record. This brings up the issue of side length, one of the most important aspects of helping the final record to sound beautiful.

Depending on the level and density of the material, a rule of thumb is to keep a 33⅓ rpm 12-inch LP at less than twenty minutes a side in length. Sixteen to eighteen minutes is ideal to make it sound really good. For a 45 rpm 7-inch single, try to keep the sides less than four minutes long. From three to three and a half minutes in length will usually sound great. In general, the shorter the side, the better it will sound.

What If the Songs Will Not Fit?

It seems to be fairly common for modern records to be longer than 40 minutes, so here are some things to think about when preparing the master for vinyl if the sides are too long. All these options entail going back to the producer and letting him or her, the band, and/or the label decide what they want to do if it won't fit.

Remove Songs
It's pretty easy to say that the most unpopular suggestion is to remove a song (or two) from the vinyl version of the album. Most bands include download codes when you buy a vinyl record, so it can be a big problem when something slated to be included as a free download has to be cut from the vinyl record.

Shorten Songs
Shorten the songs by having the mixes faded out earlier than on the CD version.

Edit
Do some creative editing to cut out sections of the arrangement that are repetitive or don't make the song better.

Change the Sequence

Change the sequence from the CD version to try to balance out the running times of the sides. Within a minute or less of each other is good practice, but if one side is more than a couple of minutes longer than the other, consider putting more of the less dynamic, less bass/low-frequency energy–filled, quieter songs on one side.

Cut It to Four Sides Instead of Two

This doubles the cost of cutting, plating, and manufacturing, but it sounds fantastic. If the record is more than 43 minutes long with shortened fades, edits, and lowered levels, the principals should think about going for four sides, putting it all back in, and getting a higher-fidelity result. If the side lengths fall into the right range, the LP can be cut at 45 rpm for even better fidelity, which usually sounds fantastic.

Processing for Master Lacquer

When I am cutting vinyl, making what is called a *direct transfer*, there are about five main things I can do with processing to get the master lacquer cut with the best potential sound quality. All these processes are subjective.

High-Frequency Limiting

This is usually either bypassed or set in a minimal way to catch high-frequency transients that could cause a skip or damage the cutter head.

Low-Pass and High-Pass Filtering

Filtering at 40 Hz and below is usually a great starting place. If the cutter is having trouble fitting the material on a side, rolling off more low end can be a big space saver.

Level Adjustments

This is probably the main thing a cutter can do to save space if there is trouble fitting the material on a side.

Elliptical EQ

This is a crossover that moves all low frequencies below a preset frequency to the center. Elliptical EQs are used this way to stop excessive lateral movement of the stylus if there is too much low-frequency energy on one side. This happens, for instance, when the kick drum, bass guitar, or bass synthesizer is panned to one side.

De-Essing

I prefer not to have to do this by the time it gets to me. If there is a sibilance problem, I feel that it is better and more specifically handled by the mastering engineer and even more so by the mixer. In other words, it's a shame to have to de-ess your whole mix because, for example, the hi-hat is too loud. It takes other good stuff in that frequency range down with it.

Cutting the Master Lacquer

With all this in mind, let's go through the process of cutting the master lacquer. First, I listen through the material and then figure out how to best set up the lathe for the best sound within the physical limits mentioned previously. I then do a simulation run on the lathe, letting it run through the side, manually banding between songs—all without actually dropping the cutter head and making a cut. These are called *dummy runs*. I do this and then make adjustments to the processing until I can get the recordings to fit, sounding the best it can. I also usually do a test cut on scrap lacquer and listen back to it. I often skip around and cut passages that I think are potential trouble spots and see how they play back before I commit. Once this happens, I bring out a blank master lacquer from its dust-free box. I visually inspect it for flaws, and if I do not see any, I put it on the turntable, turn on the suction, and cut the master lacquer.

The master lacquer cannot be played once cut. I inspect it with a microscope for overcuts or flaws. If there are none, the last thing to do is hand scribe the matrix number and the side info in the space between the label and the run-out groove and then immediately box it up and ship it out to the plating/pressing plant. Clients should be sure that their order and contact information have been received correctly with the plant before the master lacquer is cut. This is because a master lacquer degrades if it sits in a box for days. Clients should be made aware of this as early as possible. The cutting should not start until the shipping and ID information is given to the vinyl cutter so that shipping can happen immediately.

Weeks go by, the test pressing comes in the mail, and you listen.

David A. Hoatson is the cofounder and Chief Software Engineer of Lynx Studio Technology, Inc. Lynx equipment is considered to be of the highest quality, with several popular audio interfaces used for professional mastering.

This contribution helps us to understand the important differences between using ASIO and WDM drivers.

ASIO versus WDM

David A. Hoatson, Cofounder and Chief Software Engineer
Lynx Studio Technology, Inc.

History and Implementation

ASIO is an acronym for Audio Stream Input/Output, which was developed primarily by Stefan Scheffler of Steinberg (now owned by Yamaha) starting around 1997. It is primarily a user-mode application programming interface (API), meaning part of the audio card driver is loaded by the application and uses the same address space as the application. This also means that the driver must be written in (at least) two parts—the user-mode component that is loaded by the application and the kernel-mode component that is used to talk to the audio hardware. I say "at least" two parts because in 64-bit environments there must be a 64-bit kernel-mode driver, a 64-bit user-mode driver for native 64-bit applications, and a 32-bit user-mode driver for 32-bit applications. ASIO is currently supported only on Windows operating systems. Originally it was supported on Mac OS 9 as well, but it was never ported to OS X (although there are no technical reasons why it couldn't be).

ASIO provides a direct connection between the application's audio engine and the audio hardware. For most implementations, hardware manufactures allow the audio application to write directly into the buffer that the audio hardware is accessing to play and record the audio. This means that there are no intermediate processes that might affect audio quality or processor performance. ASIO is implemented as a Ping-Pong buffer, where buffer A is being filled (or emptied) by the hardware, and buffer B is being processed by the application. Once the hardware and driver are done with buffer A, it signals the application (using an interrupt and/or event) that it is switching to buffer B, and the application can start servicing buffer A. If done properly, the application has the maximum amount of time to process the audio buffer before it becomes active by the hardware. Glitches can occur in the audio when the application doesn't have enough time to process the buffer before that buffer becomes active. Reasons for this are varied: too many active plug-ins, too small a buffer size, and other hardware or software stealing the processor away and preventing the audio application from completing its task, just to name a few.

WDM is an acronym for Windows Driver Model, which was developed by Microsoft starting around 1997 to unify drivers between Windows 98 and Windows 2000. Contrary to popular belief, WDM drivers actually present no API directly to applications. All calls from the application must use the API from Microsoft Multimedia Extensions (MMEs or waveOut); DirectSound, or Direct Kernel Streaming. With the introduction of Windows Vista, Microsoft added a fourth API to replace Direct Kernel Streaming: WASAPI. Windows then interprets these calls into a single API that is sent down to the kernel-mode WDM audio driver stack and on to the audio driver. The audio is reprocessed and may be mixed or sample rate-converted by the Microsoft code.

Originally, there were two types of WDM drivers: WaveCyclic and WavePCI. WaveCyclic was intended for use with audio hardware that uses transfers from a fixed location in memory. WavePCI was intended for use with audio hardware that can perform scatter/gather transfers to or from any location in memory. With Windows Vista and Windows 7, Microsoft added WaveRT (wave real time) that works much more like ASIO.

WDM (when used with MME or DirectSound) always has an intermediate layer provided by Microsoft that affects performance in terms of both audio quality and processor usage. There are few applications that use Direct Kernel Streaming (Cakewalk SONAR is the only one that comes to mind), which can bypass Microsoft's audio layer when using WDM. With the addition of WASAPI in Windows Vista and beyond, Microsoft has allowed application developers to more easily directly access the audio buffer that the hardware is accessing to play and record audio. This also requires a change in the driver to use the WaveRT model. Along with WaveRT came a way of allowing the driver to signal when it needed service instead of relying on the application and/or operating system to poll the driver to determine when service was required. When the application asks the driver to allocate the audio buffer with notifications, it greatly improves the reliability of audio playback using the Windows audio subsystem.

Sonic Differences

The audio driver and hardware's job is to transfer the audio from the application to the analog domain in the most efficient and transparent way possible. In a perfect world, the driver model used should have no effect on audio quality or computer performance. Unfortunately, this is not always the case. This is why ASIO has dominated the professional audio market over the past decade. It provides a low-latency way to transfer audio from the application to the audio hardware in a very efficient manner. It also keeps other parts of the operating system from altering the audio stream.

WDM has been plagued with different issues in each version of the operating system. Windows 2000 and Windows XP both suffered with KMixer touching each sample as it passed through the system, sometimes with horrible sonic consequences. Certain versions of Windows Vista suffered from truncation issues, where the audio subsystem would be given 24-bit audio but would truncate it to 16 bits (no dither) even though the audio card could play 24 bits. Windows 7 with WASAPI in exclusive mode and a WaveRT driver is really the first real alternative that promises audio performance similar to that of ASIO. The main advantage that ASIO has is its mono-channel structure. With a multichannel audio device, ASIO treats each channel as its own independent mono channel. If an application wants to do eight-channel operation, it sees eight mono channels. If it then wants to do 16-channel operation, it sees 16 mono channels. With a WDM multichannel device, the channels are interleaved into a single stream. This means that going from 8 to 16 channels causes the device to need to be switched into 16-channel mode at the operating system level, which may not be possible (Windows only knows about 7.1) or be very difficult to achieve. Windows 7 still suffers from certain bugs that affect the audio quality when using WDM multichannel devices. Specifically, when in 16-bit multichannel mode and playing back a stereo file, there is some audio being played back (at a very low level) on channels that should not be active.

Zero-Latency Monitoring

ASIO provides a feature called *direct hardware monitoring* that allows applications to directly control the feed through audio (from input to output) in a standardized way. This allows end users to set up custom mixes for headphones or studio monitors while recording without the latency caused by going through the record and playback buffers. Cubase, Nuendo, Samplitude, Sequoia, and Reaper all support direct hardware monitoring.

Although WDM provides a topology driver that could provide a similar feature set to direct hardware monitoring, each audio card manufacturer is free to implement the driver however it desires. This means that there is no standardized way for the audio application to present zero-latency monitoring using WDM. The end user is left with monitoring through the record and playback buffers, which may add several milliseconds of delay to the headphone mix. Many performers find this unacceptable.

APPENDIX A
Decibel Units
of Measure

These are the most typically used decibel suffixes and other loudness units in audio engineering:

dB   Decibels A weighting, which usually refers to dBSPL A weighting, although it could mean dBFS A weighting; it requires another modifier for specific descriptions (e.g., dBSPL A).

dBB   Decibels B weighting, which usually refers to dBSPL B weighting, although it could mean dBFS B weighting; it requires another modifier for specific descriptions (e.g., dBSPL B).

dBC   Decibels C weighting, which usually refers to dBSPL C weighting, although it could mean dBFS C weighting; it requires another modifier for specific descriptions (e.g. dBSPL C).

dBFS   Decibels relating to full scale, a description of signal intensity in the digital domain.

dBRMS   Decibels root mean square, the average decibel level; for specific descriptions, it requires another modifier (e.g., dBSPL RMS).

dBSPL   Decibels sound-pressure level, a description of loudness in the real world.

dBTP   Decibels true peak.

LKFS   Loudness, K-weighted, relative to full scale.

LU   Loudness units.

LUFS   Loudness units relative to full scale.

Mastering Resources and References

C ollecting and keeping up with the resources below can help to grow knowledge of audio mastering. Keeping up with the forums and reading posts from the past can give deep perceptions about mastering. The Berklee School of Music course, YouTube videos, and the DVD are great for hearing the actual sound samples they provide and giving very clear insight into the principles. RSS feeds are available on some of the online sources, which can help to alert a faithful mastering engineer about new information. Social media is another powerful tool for getting to know other mastering engineers.

Online Resources

- *Gearslutz.* The one and only Gearslutz forum! It is the most active forum to date and has an enormous wealth of information about mastering: www.gearslutz.com/

- *Dave Collins Forum at Pro Recording Workshop* http://prorecordingworkshop.lefora.com/

- *Square Cad* http://squarecad.net/

- *Pensado's Place* www.pensadosplace.tv/

- *Tape Op Magazine* www.tapeop.com/

- *Recording.org*
  www.recording.org/

- *Prosoundweb*
  www.prosoundweb.com/

- *Soundstrips*
  www.soundstrips.com/

- *KVR*
  www.kvraudio.com/

- *YouTube*. Search audio mastering–related keywords.

- *Bobby Owsinski's Blog*
  http://bobbyowsinski.blogspot.com/

- *Paul Grundman's Mastering Matters*
  www.prosoundnetwork.com/Default.aspx?tabid=69&blogid=16

- *Mastering Tuition*
  www.masteringtuition.com/

- *AES*. Access hundreds of documents. Also, there is the open AES Oral History Project online:
  www.aes.org

- *Wikipedia*. As with many subjects, Wikipedia provides a wealth of information and points to sources that can provide even more detail, when needed.

Offline Resources

Print

- Anderton, Craig. *Audio Mastering*. New York: Peter Gorges, 2002 (out of print).

- Boden, Larry. *Basic Disc Mastering*, 2nd ed. Cincinnati, OH: Larry Boden, 2012.

- Cousins, Mark and Heppworth-Sawyer, Russ. *Practical Mastering: A Guide to Mastering in the Modern Studio*. Burlington, MA: Focal Press, 2013.

- Gallagher, Mitch. *Mastering Music at Home*. Shasta Lake, CA: Artistpro, 2007.

- Gibson, Bill. *Mixing and Mastering.* Boston: Thomson Course Technology, 2006.

- Katz, Bob. *iTunes Music.* Burlington, MA: Focal Press, 2013.

- Katz, Bob. *Mastering Audio: The Art and the Science*, 2nd ed. Burlington, MA: Focal Press, 2007.

- Orfanidis, Sophocles, J. *Introduction to Signal Processing.* Saddle River, NJ: Prentice Hall, Inc., 1996.

- Owsinski, Bobby. *The Audio Mastering Handbook.* Boston: Thomson Course Technology, 2008.

- Owsinski, Bobby. *Mixing and Mastering with IK Multimedia T-RackS: The Official Guide*, 1st ed. Independence, KY: Course Technology, 2010.

- Pettengale, Paul. *MusicTech Focus Mastering*, Vols. 1, 2, and 3. London: Anthem Publishing, 2011.

- Pohlmann, Ken C. *The Compact Disc Handbook.* Madison, WI: A-R Editions, 1992.

- Pohlmann, Ken C. *Principles of Digital Audio.* New York: McGraw-Hill, 2011.

- Rose, Jay. *Audio Postproduction for Film and Video.* Burlington, MA: Focal Press, 2009.

- Tischmeyer, Friedemann. *Audio-Mastering mit PC-Workstations.* Santa Cruz, CA: Wizoo, 2006 (in German).

- White, Paul. *Basic Mastering* (The Basic Series). New York: Music Sales America, 2006.

Video

- *Mastering Revealed.* Vancouver, BC: Streamworks Audio, 2010.

- Tischmeyer, Friedemann. *Friedemann Tischmeyer: Audio Mastering Tutorial*, DVD Vols. 1 through 3. Santa Cruz, CA: Tischmeyer Publishing, 2007.

Classes

- Berklee School of Music. Audio Mastering Techniques.
- Alchemea Mastering Techniques, Weekend Course on Mastering: www.alchemea.com
- New York Institute of Forensic Audio

Index

3D (three-dimensional) sound, dealing with lack of, 108

–3dBFS
    receiving/importing recordings, 19
    setting peaks at, 13

5.1 audio
    home theatre systems, 152
    mastering for, 49–50, 148
    monitors (speakers), 29

16-bit format
    gain staging and, 106
    receiving/importing recordings, 19

24-bit format
    gain staging and, 106
    receiving/importing recordings, 19

32-bit format
    gain staging and, 106–107
    software plug-ins and, 26

64-bit architecture, 26

1394 (FireWire)
    audio interfaces, 158–159
    connections, 54

A

AAC (Advanced Audio Coding), 147

absorption, in acoustics, 58

absorption coefficients, 58

acoustics
    absorption and diffusion, 58
    acoustic environment simulation for headphones, 47
    aspects of mastering sessions, 4–5
    dance clubs, 152
    devices for, 27–28
    equalization, 62
    hiring professional in, 57
    noise minimization, 59
    room dimensions and, 61–62
    speaker and subwoofer placement, 63–66
    techniques/design concepts, 66–67
    translation and, 66
    treatments, 58–61

active monitors, vs. passive monitors, 28–29

A/D (analog-to-digital)
    clipping A/D converter, 122
    converters, 32–33

DSPs (digital signaling processors), 185–186
    raising levels prior to analog processing, 98

ADAT Lightpipe optical connections, 54

adjustments/improvements
    comparing in bypass mode, 74
    making at same level of loudness, 74–76
    minimizing delay between comparisons, 77–78
    passes during mastering sessions, 5
    revisions to output, 136
    time requirement for editing, 22–23

Advanced Audio Coding (AAC), 147

Advanced Television Systems Committee (ATSC), 124–125

AES (Audio Engineering Society), 166

AES/EBU
    audio interfaces, 34
    connections, 53
    wordclocks and, 35–36

AIFF (Audio Interchange File Format), 145

"air" frequencies
    adding noise for brilliance, 99
    balancing, 86
    balancing with bass frequencies, 16

albums
    clearinghouse for album credit information, 150
    sales at all time low, 162
    sequencing songs on, 129

algorithms, for differing, 95

AllMusic Database, 150

AM radio, radio processing, 151

amplifiers
    quality of, 67
    types of, 31–32

analog
    analog-to-digital. *See* A/D (analog-to-digital)
    components, 26
    compressors, 39–40, 117–118
    dealing with harshness of sound, 104
    de-essing, 44
    digital-to-analog. *See* D/A (digital-to-analog)
    equalizers, 36–37, 83

analog (*continued*)
    gain staging and, 107
    harmonic distortion and, 182
    harmonic enhancement and saturation, 45
    Jaakko Viitalähde on connection and calibration of analog mastering chain, 175–180
    locating hum in analog chain, 97–98
    meters, 50
    M/S (mid-side) processors, 43
    multiband compression, 41, 169–172
    raising levels prior to analog processing, 98
    receiving/importing recordings, 20
    software plug-ins vs. analog processing, 26–27
    XLR cables, 52
antivirus software, disabling during mastering, 158
archiving DDP files, 142
ASIO (Audio Stream Input/Output), 198–200
asymmetrical waveforms, 96
ATSC (Advanced Television Systems Committee), 124–125
attack/release threshold settings, 113–114
attended sessions
    receiving/importing recordings and, 20–21
    vs. unattended, 4
Audio Engineering Society. *See* AES (Audio Engineering Society)
audio forensics, 167
Audio Interchange File Format (AIFF), 145
audio interfaces
    FireWire (1394), 158–159
    overview of, 34–35
Audio Stream Input/Output (ASIO), 198–200
autos
    quality of recordings in, 109
    speaker placement in, 151

B

backup options, for DAWs, 154–155
badware, 159–160
Bantam/TT (Tiny Telephone) connections, 54
bass
    balancing bass frequencies, 84–85, 100–101
    balancing with "air" frequencies, 16
    dealing with harshness of digital sound, 103
    enhancing bass frequencies, 96
bass traps, 61

Baxandall shelves, 89
BD (Blu-ray Disc), 146–147
Bessel filters, 88
BIN files, as alternative to DDP, 142–143
BIOS, disabling onboard sound, 158
bit rate
    meters, 132
    sample range and, 11
biwire connections, between amplifiers and speakers, 67
Blackwood, Brad
    bio section on, 184
    on Mid-Side processing, 184–185
Blue Book standards, 137
Blumlein shuffling, 42
Blu-ray Disc (BD), 146–147
BNC connections, 54
boundary effects, in speaker placement, 64
brilliance, adding noise in "air" band for, 99
broadcast processors, 192
broadcast standards, 124–126
Broadcast WAV (BWAV), 145
buffering issues, DAWs (digital audio workstations), 159
Butterworth filters, 87–88, 189
BWAV (Broadcast WAV), 145
bypass mode, listening in, 79–80

C

cabling, connections and, 52–55
CALM (Commercial Advertisement Loudness Mitigation), 124–126
cars. *See* autos
CDs
    basic processing, 7–10
    comparing processed version with original, 10–11
    DAW (digital audio workstation) and, 8
    disc life of, 151
    dithering, 11–12
    enhanced, 148
    error levels, 141–142
    exporting, 12
    issues with, 7–8
    mastering plug-in packages for, 8–9
    mastering processes for, 1
    overview of, 7
    pause length between tracks, 137–138
    premaster CDs, 140
    refractive index levels, 151–152
    replication vs. duplication, 150–151
    sample rates, 11
    track offsets, 138
CD-Text, 20, 139–140

ceilings (audio), basic limiting and, 9–10
ceilings (physical), acoustics and, 59
changes. *See also* adjustments/improvements
    making client requested changes
        promptly, 79
    making revisions to output, 136
channels
    balancing, 100
    correlation meters, 133
    monitoring, 90
characters
    tonality and, 115
    using same character on all songs, 94
Chebyshev filters, 88, 189
checksums
    checking data integrity, 20
    final output and, 144–145
clicks, noise reduction and, 129
clients
    customer service and, 22
    importance of timeliness to, 21
    making client requested changes
        promptly, 79
    previewing final output, 135–136
    requesting information from, 20
clipping
    A/D converter, 122
    loudness and, 121
clocks, 35–36
clouds, acoustic absorbers, 59
coloration
    Dave Hill on, 180–184
    understanding, 92–93
Commercial Advertisement Loudness
    Mitigation (CALM), 124–126
competition, challenges facing new mastering
    studio, 161
compression
    adding warmth, 104
    analog compressors, 117–118
    basic mastering and, 9
    equalization following, 90
    equipment for, 38–41
    file compression, 22
    input and output, 116–117
    before limiting, 121
    linked and unlinked, 117
    as mastering process, 1
    processing tools for, 5
    pros/cons of requesting removal of
        mixbus processing, 14
    settings and meters, 113
    types of compressors, 114–117
computers, audio playback quality on, 152
concurrent processing, 80

connections
    balanced to unbalanced, 100
    types of, 52–55
consoles
    acoustics and, 60
    for mastering, 46–47
    patching order and, 176
controls, stepped vs. continuously variable, 26
converters
    to/from analog and digital, 32–33,
        185–187
    clipping A/D converter, 122
    sample-rate converters, 49
correlation meters, for comparing left/right
    channels, 133
courses, for getting started in mastering, 165
CRC (cyclic redundancy checks), 20
cross-fades, 128
crossovers, monitors (speakers), 30
CUE files, as alternative to DDP, 142–143
customer service
    challenges facing new mastering studio,
        162
    receiving/importing recordings and, 22
cyclic redundancy checks (CRC), 20

D

D/A (digital-to-analog)
    converters, 32–33
    DSPs and, 186–187
    intersample peaks and, 99–100
    raising levels prior to analog processing,
        98
daisy-chaining (point-to-point wiring)
    patching method, 177–178
dance clubs, acoustics and PA systems in, 152
DAO (disk-at-once), vs. track-at-once write
    method, 140
data integrity, checking, 20
databases, submitting information to online,
    149–150
DAW (digital audio workstation)
    A/B comparisons, 176
    backup options, 154–155
    basic mastering, 8
    displaying sample rate, 11
    dithering feature of, 12
    driver issues, 156–157
    fades and cross-fades and, 128
    hardware issues, 157–158
    hardware options, 154
    maintaining, 157
    not letting dictate workflow, 153
    optimizing, 157–160

DAW (digital audio workstation) (*continued*)
    plug-in issues, 157
    receiving/importing analog recordings,
      20
    spacing songs and, 127–128
    stopping nonessential processes and
      services, 155
    techniques and functions for working
      with, 154
    types of, 47–48
    user account settings, 155–156
dB (decibels)
    monitor calibration, 68
    units of measure, 201–202
dBFS (decibels relative to full scale)
    digital limiting and, 123
    monitor calibration, 68
    receiving/importing recordings, 19
    setting peaks at -3dB, 13
    units of measure, 201
dBSPL (decibel sound-pressure level)
    monitor calibration, 68
    units of measure, 201
DC offset, 96
DDP (Description Protocol)
    final output options, 135, 142–143
    manufacturing CDs, 1
    output of mastering process, 5
    software, 48–49
deadlines, allowing time for mastering, 15
decibel sound-pressure level (dBSPL)
    monitor calibration, 68
    units of measure, 201
decibels (dB)
    monitor calibration, 68
    units of measure, 201–202
decibels relative to full scale. *See* dBFS
    (decibels relative to full scale)
de-essing
    analog and digital, 44
    dealing with harshness of digital sound,
      104
    premastering for vinyl and, 197
Deferred Procedure Call (DPC), Latency
    Checker, 157
depth/three dimensional sound, dealing with
    lack of, 108
design concepts, acoustics, 66–67
desks, acoustics and, 60
destructive processing, vs. nondestructive,
    76
detents, potentiometers and, 26
device drivers
    driver system options, 159
    overview of, 156–157

diffusers
    acoustics and, 58
    rear diffusers, 60
digital
    to analog. *See* D/A (digital-to-analog)
    analog-to-digital. *See* A/D (analog-to-
      digital)
    clipping, 121
    de-essing, 44
    equalizers, 83
    files. *See* files
    gain staging and, 107
    harmonic enhancement and saturation, 45
    limiters, 102, 121–123
    meters, 50
    multiband compressors, 41
    stereo processors, 43–44
    storing digital files, 149
digital audio workstation. *See* DAW (digital
    audio workstation)
digital signaling processors. *See* DSPs (digital
    signaling processors)
direct hardware monitoring, ASIO feature, 200
disk drives
    for DAWs, 155
    using second drive for audio, 158
disk-at-once (DAO), vs. track-at-once write
    method, 140
distortion
    Dave Hill on, 180–184
    limiting, 99
    quality loss due to loudness, 120
    technique for dealing with, 93
    understanding, 92–93
distribution amps, 35–36
dithering
    basic mastering and, 11–12
    technique for use of, 94–95
DIY (Do-It-Yourself) mastering, 162
DMG EQuality plug-in, 8
downward compression, 112. *See also*
    compression
DPC (Deferred Procedure Call), Latency
    Checker, 157
DSPs (digital signaling processors)
    A/D (analog-to-digital) converters,
      185–186
    D/A (digital-to-analog) converters,
      186–187
    for mastering, 51
dummy runs, 197
DVDs, 146
dynamic equalizers, 38, 103–104
dynamics processing
    analog compressors, 117–118

attack/release threshold settings, 113–114
compression and limiting and, 5
compression settings and meters, 113
compressor input and output, 116–117
compressor types, 114–117
expansion, 117
gain-reduction meters, 116
linked and unlinked compression, 117
macrodynamics and microdynamics, 114
RMS sensing and peak-sensing, 114
types of, 111–113
volume automation, 116

— **E** —

EAN (European Article Number), 138–139
ear fatigue, avoiding, 78
ear sensitivity, to loudness, 71, 122, 124
early reflections, acoustics and, 60
EBU (European Broadcast Union), 124–125
Eclipse systems, for replication of CDs, 150–151
editing. *See* adjustments/improvements
effects, multieffect processors, 51
Elliptical filters, 189, 196
emulator plug-ins, 93
enhanced CDs, 148
equalization/equalizers
     acoustics and, 62
     analog equalizers, 36–37
     basic mastering and, 8
     dynamic equalizers, 38
     filters and, 88–89
     frequency balancing with, 84–87
     as mastering process, 1
     matching equalizers, 102–103
     order of frequency balancing, 88
     parametric, analog, and digital equalizers, 83–84
     plug-in equalizers, 37–38
     processing tools, 5
     pros/cons of requesting removal of mixbus processing, 14
     subtractive, 88
     testing different settings, 173
     uses of equalizers, 82–83
     using following compression, 90
equipment
     5.1 mastering, 49–50
     acoustic devices, 27–28
     amplifiers, 31–32
     analog components, 26
     audio interfaces, 34–35
     clocks and distribution amps, 35–36
     compression and expansion, 38–41

connections and cabling, 52–55
consoles, 46–47
controls, 26
converters, 32–33
DAW (digital audio workstation), 47–48
DDP software, 48–49
demos, 25
DSPs (digital signaling processors), 51
equalizers, 36–38
forensic audio software, 52
harmonic enhancement and saturation, 45
headphones, 47
limiters, 42
metadata embedding software, 51
meters, 50
monitoring control systems, 28
monitors (speakers), 28–31
multieffect processors, 51
playback systems, 48
restoration and noise reduction, 44–45
retailers, auctions, and classified ads, 27
routers and patchbays, 45–46
sample-rate converters, 49
software plug-ins vs. analog processing, 26–27
stereo/mid-side processors, 42–44
subwoofers, 31
errors
     error sheets, 5
     final output and, 140–142
European Article Number (EAN), 138–139
European Broadcast Union (EBU), 124–125
expansion
     before compression, 117
     loudness and, 38–41
     technique for dealing with lack of depth, 108
exporting, 12

— **F** —

fades, 128–129
fast Fourier transforms (FFT), 131–132, 192
fast/medium transient sounds, dealing with sounds that stick out too much, 106
Fastsum, checking data integrity, 20
FET (field-effect transistor) compressors, 118
FFT (fast Fourier transforms), 131–132, 192
field-effect transistor (FET) compressors, 118
files
     file compression and file transfer, 22
     organizing, 21
filters
     equalization, 88–89
     Pieter Stenekes on, 185–191

filters (*continued*)
    premastering for vinyl and, 196
    types of, 187–190
FIR (finite impulse response), 189–190
FireWire (1394)
    audio interfaces, 158–159
    connections, 54
FLAC (Free Lossless Audio Codec), 145
FM radio, radio processing, 151
forensics
    forensic audio software, 52
    opportunities in audio forensics, 167
format conversions, audio and video over
    Internet, 152
Fourier transforms, 131–132
Free Lossless Audio Codec (FLAC), 145
frequency balancing
    balancing bass with "air" frequencies, 16
    dealing with harshness of digital sound,
      103
    distortion and color and, 181
    equalization, 84–87
    frequency spectrum and, 105
    order of, 88
    roll-off on both ends, 90
frequency spectrum, 105
front-wall, acoustic treatment for, 58–59
fullness, of sound, 105
full-range monitors, 29
furniture/objects, acoustics and, 61

G

gain
    basic limiting, 9
    clipping to raise, 122
    gain-reduction meters, 116
    makeup gain, 116
    staging, 106–107
    stepped monitor gain control, 70
gClip, 121
Gerzon shelves, 89
Gibbs phenomenon, 183
Gould, Cornelius
    bio section on, 191
    on optimizing audio for radio, 191–193
Grammy Recording Academy, 166
grounds/grounding, hum and, 97–98, 178–179

H

hard-disk drives (HDDs), 155
harmonics
    distortion and color and, 181
    equipment for harmonic enhancement, 45
harshness of digital sound, dealing with,
    103–104
HDDs (hard-disk drives), 155
headphones
    for mastering, 47
    monitoring and, 72
hearing, sample rates and, 16
high frequencies
    balancing, 86
    dealing with harshness of digital sound,
      103–104
Hill, Dave
    bio section on, 180
    on distortions and coloring, 180–184
hiss, noise reduction and, 129
Hoatson, David A.
    on ASIO vs. WDM, 198–200
    bio section on, 198
home theatre systems, 152
hotkeys, 154
Hull, Scott
    bio section on, 172
    on full-range monitoring, 172–174
hum
    grounds/grounding and, 178–179
    locating in analog chain, 97–98

I

IEEE 1394 (FireWire)
    audio interfaces, 158–159
    connections, 54
IIR (infinite impulse response) filters, 187–189
impedance, 179
importing recordings. *See* receiving/
   importing recordings
impulse response
    FIR (finite impulse response) filters,
      189–190
    IIR (infinite impulse response) filters,
      187–189
    overview of, 187
infinite impulse response (IIR) filters, 187–189
International Organization for
    Standardization (ISO), 59
International Standard Recording Codes
    (ISRC), 138
International Telecommunications Union
    (ITU)
    loudness standards, 124–125
    speaker placement guidelines, 63
Internet
    exporting MP3s to, 12
    streaming and format conversions, 152
internships, 165

intersample peaks (ISPs), 99–100
ISO (International Organization for Standardization), 59
ISPs (intersample peaks), 99–100
ISRC (International Standard Recording Codes), 138
ITU (International Telecommunications Union)
    loudness standards, 124–125
    speaker placement guidelines, 63
iTunes, mastering for, 147
iZotope Ozone, mastering plug-in package, 8–9

J

jackfields. *See* patchbays
jitter
    inaccurate clocks producing, 35
    technique for dealing with, 95

K

keyboard shortcuts, 154
K-Meter, 69
K-System
    criticisms of, 70
    K-20 (film), K-14 (rock), K-12 (broadcast and pop), 69
    monitor calibration, 68

L

latency issues
    DAW (digital audio workstation), 159
    DPC Latency Checker, 157
layback, tape machines and, 109
learn feature, noise reduction and, 129
legal liability, challenges facing new mastering studio, 163–164
levels
    making adjustments at same level of loudness, 74–76
    premastering for vinyl and, 196
    reducing before part changes, 106
    reference level. *See* reference level
limiting/limiters
    basic mastering and, 9
    digital, 42, 102, 121–122
    dithering and, 12
    equipment for, 42
    loudness, 121–123
    mastering processes, 1
    mixing into limiters, 14
    operating limiters, 122

    premastering for vinyl and, 196
    processing tools, 5
    serial limiters, 123
    as type of compression, 111, 113
linked compression, 117
listening
    in bypass mode, 79–80
    skill needed for audio engineering, 74
lossy output, 146
loudness
    apparent/perceived, 119
    basic limiting, 9–10
    broadcast standards, 124–126
    clipping and, 121
    compression and expansion, 38–41
    ear sensitivity and, 71, 122
    ideal level of, 124
    issues with basic mastering, 7
    limiting and, 121–123
    loudness war, 120
    making adjustments at same level of, 74–76
    monitoring and, 28, 70
    peaks, 124
    processing based on loudest passages, 124
    recording potential for, 121
    units of measure, 201–202
    workplace safety and, 72
L/R balance, 90. *See also* channels

M

macrodynamics
    overview of, 114
    volume automation, 116
malware, 159–160
marketing, 164
master bus (mixbus)
    compressors, 115
    pros/cons of requesting removal of mixbus processing, 14
master clocks, 35
mastering engineers
    assisting with mix issues, 17
    choosing based on genre, 2–3
    role of, 2
mastering overview, 1–3
mastering sessions, 4–5
mastering software bundles, 27
mastering studios
    benefits of, 164–165
    challenges facing, 161–164
    help in getting started, 165–167
    hiring interns, 167
    opportunities in audio forensics, 167

mastering studios (*continued*)
   overview of, 161
   related fields, 167–168
maximizers, in limiting, 42, 123
MCNs (Media Catalog Numbers), 138–139
MD5 (Message-Digest Algorithm), 20, 144–145
metadata
   embedding software, 51
   MP3s and, 20
   nonlossy, 145
meters
   action and speed of meters, 134
   analog and digital, 50
   bit meters, 132
   correlation meters, 133
   FFT/Fourier transforms, 131–132
   overview of, 131
   reconstruction meters, 134
   spectral analyzers, 132–133
   vectorscope, 133–134
mice, input options for DAWs, 154
microdynamics
   overview of, 114
   volume automation, 116
middle channels, monitoring, 90
middle signal, M/S (mid-side) processors, 42
midfield monitors, 29
midrange frequencies
   balancing, 85–86
   dealing with harshness of digital sound,
      103
Mid-Side processing
   Brad Blackwood on, 184–185
   overview of Mid-Side processors,
      42–43
   technique for, 91
   using unique Mid-Side processors, 92
minijack connections, 53
mixbus (master bus)
   compressors, 115
   pros/cons of requesting removal of
      mixbus processing, 14
mixes/mixing
   balancing bass and "air" frequencies, 16
   dealing with bad mixes, 97
   fades during, 128
   limiting, 13–14
   mixbus processing and, 14
   multiple mono vs. stereo interleaved, 14
   not performing mastering on own
      mixes, 3
   overview of, 13
   problems with, 16
   sample rates, 15–16
   stages of production, 1

stem mastering, 14–15
tape and, 16
working relationship with mixing
   engineers, 80
monitoring
   aspects of mastering sessions, 5
   control systems for, 28
   headphones and, 72
   loudness levels and, 70
   middle and side channels, 90
   Scott Hull on full-range monitoring,
      172–174
   zero-latency monitoring, 200
monitoring D/A converter, 32
monitors (speakers)
   5.1 systems, 29
   acoustics and, 62
   active vs. passive, 28–29
   amplifiers, 31–32
   biwire connections, 67
   calibration, 68
   crossovers, 30
   decoupling to eliminate sympathetic
      resonances, 61, 64
   full-range, midfield, and near-field, 29
   in mastering sessions, 5
   placement of, 30–31, 63–66
   selecting, 67
   stepped monitor gain control, 70
   types of, 28–31
   why small monitors are not ideal, 173
mono
   cables, 53
   compatibility checks, 91–92
   stereo interleaved vs. multiple mono, 14
MP3s
   audio playback quality with MP3
      players, 152
   encoding without other processing, 8
   exporting to Internet, 12
   lossy output, 146
   mastering processes for, 1
   metadata, 20
   output options, 5, 135
MPEG-1/MPEG-2 Audio Layer III. *See* MP3s
M/S (mid-side) processors. *See* Mid-Side
   processing
muddiness, of sound, 104
multiband compression
   applying compression to specific
      frequencies, 41, 111, 113
   Robin Schmidt on, 169–172
multiband limiters, 123
multiband processors, 97
multieffect processors, 51

multi-pin connectors, 53
multiple mono, vs. stereo interleaved, 14

N

NAMM (National Association of Music
    Merchants), 166
native processing, 27
NC (noise criterion), measuring noise levels,
    59
near-field monitors, 29
Niveau/Tilt Filter, 89
noise
        equipment for reducing, 44–45
        minimizing, 59
        out-of-band noise, 86–87
        reducing, 129–130
noise criterion (NC), measuring noise levels,
    59
noise rating (NR), 59
nondestructive processing, 76
nonlossy output, 145
NR (noise rating), 59
NyquistShannon sampling theorem, 16

O

Omnia radio broadcast processing, 151
online backup, for DAWs, 154
Opto/ELOP analog compressors, 117–118
Orban radio broadcast processing, 151
out-of-band noise, 86–87
output
        5.1 audio and, 148
        aspects of mastering sessions, 5
        CD-Text and, 139–140
        checksums applied to, 144–145
        client preview of, 135–136
        as DDP files, 142–143
        as enhanced CDs, 148
        error checking and error levels,
            140–142
        international recording codes and,
            138–139
        in iTunes mastering, 147
        lossy output, 146
        media options for, 146–147
        nonlossy output, 145
        overview of, 135
        pause length between tracks, 137–138
        PQ sheets, 144–145
        as premaster CDs, 140
        quality control and standards, 136–137
        revising, 136
        ringtone mastering, 148
        shipping, 143–144
        track marker settings, 137
        track offsets, 138
        vinyl mastering, 143
        write method options, 140
Ozone IRC III, 14

P

parallel compression, 39, 111–112
parametric equalizers, 83–84
passband filters, 87
passes, making adjustments during mastering
        sessions, 5
passive monitors, 28–29
patchbays
        for mastering, 45–46
        patching methods, 177–178
patching, 176–178
pause length, between tracks, 137–138
PCI (Peripheral Component Interface), 55
PCM (pulse-code modulation), 32
peak program meters (PPMs), 134
peaks
        loudness, 124
        manually reducing, 106
        setting at –3dB, 13
        units of measure, 202
peak-sensing compressors, 114
Peripheral Component Interface (PCI), 55
phase rotators, in broadcast processing, 193
playback systems, 48
Plextor, error checking with, 140–141
PluginAlliance Big4 Bundle, 8–9
plug-ins
        for basic mastering, 8–9
        compressors, 41
        DAW conflicts, 157
        emulation of, 93
        equalizers, 37–38
        software plug-ins vs. analog processing,
            26–27
PMCDs. *See* premaster CDs
point-to-point wiring (daisy-chaining)
        patching method, 177–178
pops, noise reduction and, 129
potentiometers, for continuously variable
        controls, 26
Powell, Jeff
        bio section on, 194
        on premastering for vinyl, 194–197
PPMs (peak program meters), 134
PQ sheets
        album and track info on, 5
        final output and, 144–145

premaster CDs
    DDP vs., 142
    media options for output, 5, 140, 146–147
    replication vs. duplication, 150–151
processing
    based on loudest passages, 124
    for basic mastering, 7–10
    comparing versions, 10–11
    concurrent, 80
    dealing with bad mixes, 97
    dealing with harshness of digital sound,
        103–104
    destructive vs. nondestructive, 76
    dynamics processing. *See* dynamics
        processing
    native processing, 27
    raising levels prior to analog processing,
        98
    receiving/importing recordings, 22
    reverb processing, 82
    saving, copying, pasting configurations,
        21
    in stages, 80–81
    techniques, 78–79
    tools for, 5
    using same character on all songs, 94
producers, working relationship with, 80
production stages, 1
pulse-code modulation (PCM), 32
punchy compression, 115
PWM (pulse-width modulation), 39–40, 118

Q

Q values, shelves and, 87–88
QRDs (quadratic residue diffusers), 58
quality control, 136–137

R

radio processing
    AM and FM, 151
    Cornelius Gould on optimizing audio
        for, 191–193
RAID (redundant array of independent
    disks), 154
rainbow books standards, 136
RAM (random access memory), 157–158
random access memory (RAM), 157–158
RAR file compression, 22
RCA connections, 53
rear diffusers, 60
receiving/importing recordings
    analog recording, 20
    attended/unattended sessions, 20–21

CRC and checksums for checking data
        integrity, 20
    customer service and, 22
    file compression and file transfer, 22
    organizing digital files, 21
    overview of, 19–20
    processing, 22
    requesting information from clients, 20
    time requirement for editing, 22–23
reconstruction meters, 134
recording A/D converter, 33
recordings
    loudness potential of, 121
    receiving/importing. *See* receiving/
        importing recordings
    sound quality in cars, 109
    working with reference recordings, 77–78
recording/tracking stage of production, 1
Red Book
    CD error levels, 141–142
    output standards, 136–137
Red Book audio CD
    exporting WAV file to CD, 12
    output of mastering process, 5
redundant array of independent disks
    (RAID), 154
reference CDs, 135
reference level
    acoustic environments and, 66
    alternative selections, 71–72
    consistency resulting from use of, 70
reference recording
    minimizing delay between
        comparisons, 77–78
    working with, 77
refractive index levels, CDs, 151–152
registry issues, 159
reproducing (pitching) D/A converter, 33
resonance
    dealing with harshness of digital sound,
        103
    resonance processors, 96
resources
    classes, 205
    help in getting started, 165
    offline resources, 204–205
    online resources, 203–204
    video, 205
restoration equipment, 44–45
reverb processing, 82
revisions. *See* adjustments/improvements
ringtone mastering, 148
RMS (root-mean-square)
    analog compression and, 39–40
    in broadcast processing, 193

monitor calibration, 68
types of compressors, 114
room dimensions, in acoustics, 61–62
room modes, in acoustics, 57
root-mean-square. *See* RMS (root-mean-square)
routers, for mastering, 45–46
RT60/RT30/RT20, acoustic targets for mastering rooms, 66

━ **S** ━

SACD (Super Audio Compact Disc), 146
sample rates
    basic mastering and, 11
    converters, 49
    preparing mixes for mastering, 15–16
    receiving/importing recordings, 19
    technique for sample-rate conversion, 100
    upsampling, 101–102
samples, 11
saturation
    equipment for, 45
    understanding, 92–93
saving work on DAWs, early and often, 154
Schmidt, Robin
    bio section on, 169
    on multiband compression and audio gear, 169–172
self-terminating digital devices, 35
sequencing songs, of albums, 129
serial compression, 116
serial limiters, 123
sessions
    aspects of mastering sessions, 4–5
    receiving/importing recordings and, 20–21
shedding, in tape deterioration, 110
shelves. *See also* filters
    Baxandall shelves, 89
    Gerzon shelves, 89
    Q values and, 87–88
    when to use high shelf, 103
shipping final output, 143–144
shoulder frequency, in Baxandall shelves, 89
shuffling, mid-side processors and, 42
sibilance, 98
side-chain compression, 111–113
side-chain feature
    adding warmth, 104
    in compressors, 39
side channels, 90
side signal, 42

Slate Digital FGX, 14
smart phones, audio playback quality, 152
software plug-ins, vs. analog processing, 26–27
solid state drives (SSD), 155
solo modes, in seamless comparisons, 78
songs
    premastering for vinyl, 195
    processing song sections separately, 78–79
    sequencing on albums, 129
    spacing between, 127–128
    using same character on all songs, 94
songwriting stage, of production, 1
sound cards, disabling onboard sound, 158
sound pressure-level (SPL) meter, 59–60
S/PDIF
    audio interfaces, 34
    connections, 53
    wordclocks and, 35–36
speakers. *See* monitors (speakers)
spectral analyzers, for visualizations, 132–133
spectral dynamic processors, 97
spectral editing, noise reduction and, 130
SPL (sound pressure-level) meter, 59–60
spyware, 159–160
SSD (solid state drives), 155
startup costs, challenges facing new mastering studio, 163
startup settings, DAWs (digital audio workstations), 159
stems/stem mastering
    preparing mixes for mastering, 14–15
    technique for, 81
Stenekes, Pieter
    bio section on, 185
    on digital filters, 185–191
stepped controls
    on mastering gear, 26
    stepped monitor gain control, 70
stereo cables, 53
stereo interleaved, vs. multiple mono, 14
stereo processors, 43
sticky shed, in tape deterioration, 110
storage, of digital files and tape, 149
streaming audio and video, over Internet, 152
studios. *See* mastering studios
subsonic frequencies, balancing, 84
subtractive equalization, 88
subwoofers
    calibrating, 64–66
    crossovers, 30
    placement of, 64
    types of, 31

sumdifference, M/S (mid-side) processors, 42
Super Audio Compact Disc (SACD), 146
surfaces, acoustics and, 60
sympathetic resonances, 59

▬▬ **T** ▬▬

tablets, hardware options for DAWs, 154
tape
    harmonic distortion and, 181
    mastering to, 16
    mixing down to or mastering with,
        109–110
    receiving/importing analog recordings,
        20
    storage, 149
Task Manager, 155
Thunderbolt (Light Peak) connections, 54
time-domain problems, distortion and, 183
Tiny Telephone (Bantam/TT) connections, 54
tip-ring-sleeve (TRS) connectors, 53
tip-sleeve (TS) connectors, 53
tonality
    characters and, 115
    issues with basic mastering, 7
track-at-once, vs. DAO (disk-at-once) write
    method, 140
trackballs, input options for DAWs, 154
tracks
    pause length between, 137–138
    track marker settings, 137
    track offsets, 138
T-Racks mastering plug-in package, 8–9
transformers, balancing inputs and outputs
    with, 178
translation
    acoustic environments and, 66
    goals of mastering, 1–2
transparent processors, 115
TRS (tip-ring-sleeve) connectors, 53
TS (tip-sleeve) connectors, 53

▬▬ **U** ▬▬

unattended sessions
    vs. attended, 4
    receiving/importing recordings and,
        20–21
unlinked compression, 117
UPC (Universal Product Code), 138–139
upsampling, 101–102
upward expansion, 112. *See also* expansion
USB connections, 54
user account settings, 155–156

▬▬ **V** ▬▬

Variable mu compressors, 118
VCA (voltage controlled amplifiers), 39–40,
    117–118
vectorscope, 133–134
Viitalähde, Jaakko
    bio section on, 175
    on connection and calibration of analog
        mastering chain, 175–180
vinyl
    Jeff Powell on premastering for vinyl,
        194–197
    mastering for, 143
viruses, 159–160
visual media, differing and, 94
visualizations/metering. *See* meters
vocals
    dealing with lack of clarity, 102
    mastering with focus on, 107
    vocalup mix, 108
voice biometrics, in forensic audio software,
    52
voltage controlled amplifiers (VCA), 39–40,
    117–118
volume automation, in dynamics processing,
    116
Voxengo Elephant, 14

▬▬ **W** ▬▬

WAV (Waveform Audio File Format)
    exporting to CD, 12
    manufacturing CDs, 1
    nonlossy output, 145
    output options, 5, 135
    PCM (pulse-code modulation) and, 32
    receiving/importing recordings, 19
waveforms, dealing with asymmetrical, 96
Waves Masters Bundle, 8–9
WDM (Windows Driver Model), 198–200
WinMD5, 20
wordclocks, 35
workflow, not letting DAWs dictate, 153
write method options, 140

▬▬ **X** ▬▬

XLR cables, 52

▬▬ **Z** ▬▬

ZIP
    DDP files, 142
    file compression types, 22